A STORM IN THE PORT

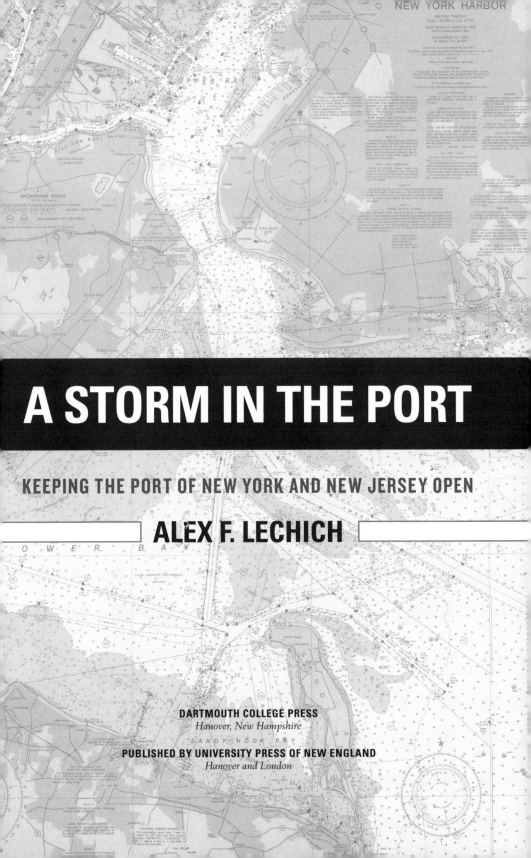

A STORM IN THE PORT

KEEPING THE PORT OF NEW YORK AND NEW JERSEY OPEN

ALEX F. LECHICH

DARTMOUTH COLLEGE PRESS
Hanover, New Hampshire

PUBLISHED BY UNIVERSITY PRESS OF NEW ENGLAND
Hanover and London

Dartmouth College Press
Published by University Press of New England,
One Court Street, Lebanon, NH 03766
www.upne.com
© 2006 by Alex F. Lechich
Printed in the United States of America

5 4 3 2 1

Library of Congress Cataloging-in-Publication Data
Lechich, Alex F.
A storm in the port : keeping the port of New York
and New Jersey open / Alex F. Lechich — 1st ed.
p.   cm.
Includes bibliographical references and index.
ISBN–13: 978–1–58465–470–4 (cloth : alk. paper)
ISBN–10: 1–58465–470–8 (cloth : alk. paper)
1. Dredging spoil—Environmental aspects—
New York Harbor (N.Y. and N.J.)
2. Dredging—Government policy—United States.
3. Port Authority of New York and New Jersey—
Management. I. Title.
TD195.D72.L45   2006
363.739'4097471—dc22   2005032140

THIS BOOK IS DEDICATED TO MY DEPARTED MOTHER, ERSILIA,
AND TO MY SON, MAX, WHO TOGETHER HAVE GIVEN ME
MY WARMEST MEMORIES AND MY BRIGHTEST HOPES.

CONTENTS

A shipping crisis serious enough to affect the entire regional economy developed in the Port of New York and New Jersey during the 1990s and, though not widely known, in many respects persists today. It was brought on by the stoppage of almost all dredging of the Port's navigational channels and shipping berths, which need to be maintained periodically for the Port to be able to function. The dredging was effectively stopped because ocean dumping of the dredged material was severely restricted by new requirements implemented by the U.S. Environmental Protection Agency (EPA). The other important factor precipitating the crisis was the almost complete lack of alternative disposal options for the dredged material. As in many other large ports around the world, millions of cubic yards of material, mostly muds and sands, need to be dredged from the Port in order to maintain the water depth that ships need to operate in the channels and berthing areas. The dredged material comes mostly from the silt suspended in the rivers that flow into the Port, partly from materials brought in with ocean tides, and partly from sources associated with the huge surrounding human population.

The massive amount of dredged material produced annually in the Port, much of it contaminated to some degree, was for many years disposed of in the nearby coastal Atlantic Ocean. The onset of new ocean disposal restrictions was predictable by at least the late 1980s, if not earlier. There had been ongoing discussions of the need for changes in the testing and evaluation of material proposed for ocean disposal among the relevant government agencies and the Port Authority of New York & New Jersey (PANYNJ). The changing environmental and public policy views regarding ocean dumping had been in the wind all through the 1980s and early 1990s. Ocean dumping of other materials had been stopped during this time, and it was common knowledge that dredged material was going to be the next focus. As far as finding alternatives to ocean disposal, it should also have been no surprise to anyone that the New York metropolitan area would be a difficult place to locate land-based disposal alternatives, which typically require substantial tracts of open land.

The crisis started taking form in the early 1990s, as environmental regulations that would make ocean disposal much more difficult were actually implemented. By the mid-'90s the backlog of dredging projects was so critical that it was feared the Port's major shipping firms would begin to leave for other ports that had better operating conditions. The Corps of Engineers and the other major federal agency involved in dredging, EPA, along with the concerned environmental groups and politicians, could not resolve their many regulatory and scientific conflicts, and brought the dredging program to a standstill. The crisis was finally mitigated through an agreement between the agencies and environmental groups brokered in July 1997 by Vice President Al Gore. Although that event may be only a milestone in a period of continuing tribulation for the Port's shipping and waterfront industries, it nonetheless marked the return to some normalcy and stability. It may also have saved the Port from a spiraling decline and perhaps even obsolescence (as will be explained just ahead). The Port industry has nevertheless been struggling since then under the twin clouds of much higher disposal costs and a continuous, almost case by case search for disposal options at those high costs. Many facilities, mainly smaller ones such as local marinas or small waterfront businesses, were unable to absorb the higher costs and have either closed or severely downsized or modified their operations.

The main players in the story were the U.S. Army Corps of Engineers, through its permitting and navigational dredging programs, and the U.S. Environmental Protection Agency, with its authority to review and veto ocean dumping activities. The other important forces at work were the public environmental groups that became involved and the Port's industrial and economic interests, including, not in the least, the Port Authority of New York & New Jersey. The development of a crisis situation such as this one is usually due at least to some degree to bad decisions, and the dredging crisis did not break this norm. Although the Corps carries the greatest burden in this regard mainly because of its historically primary role in the national and regional dredging programs, many of the other participants also have their burden to bear. This book recounts the events in environmental policy and science as they occurred, and includes the author's analysis of each group's role in the crisis.

The author was the EPA regional dredging program manager from the late 1980s to the mid-1990s, when, as the crisis came to full bloom, additional scientific staff were brought into the program to deal with the increasing workload and program management occurred essentially at successively higher levels in the agency. The program changed during that time from being a small component of the Marine & Wetlands Protection Branch, whose role in the Corps-dominated dredging/ocean disposal program was minor, to one in which EPA began to play an increasingly significant role. As the crisis became more intense, even the Regional Executive Office, meaning here the regional administrator (RA) and/or the deputy RA, became intimately involved on a day-to-day basis over a long time period. This was unprecedented, since that office's responsibilities encompass oversight of all EPA programs in the region including the other water programs, the air program, hazardous waste, etcetera. During this time, for instance, important decisions were made at the EPA Region 2 office that involved crucial issues regarding the New York City water supply. Nevertheless, the regional dredging crisis was the top priority in EPA Region 2 for far longer than any other problem.

This book was written mostly from daily logs kept by the author during most of his time at EPA. Fairly detailed notes were kept to try to keep track of the wide-ranging and continuously changing issues and events. The Region 2 dredging program covered several geographic areas, all with their separate sets of issues (although the Port was by far the gorilla in the room), and involved several different environmental science disciplines, including bioassay and chemical testing, ocean monitoring and site evaluation, and other related program areas such as decontamination technologies. This was a time of great changes in all these areas, and many of them were happening at a fast pace. At no other position held before or since has the author felt a need to keep such detailed notes, and their existence was critical in the ability even to contemplate writing the book. These and personal recollections are the basis for the book, with clarification or detailing and updating added from personal sources and from reference material.

Although the author considers himself to be quite environmentally oriented, he believes that for environmental advance to be broadly accepted in our socially and economically complex world, there is often a need to choose and conduct our battles wisely. Any activity that is targeted for regulation must have a real potential for causing adverse effects, and these must be reasonably well understood. Likewise, there needs to be realism in prioritizing targets for regulation according to their actual potential for causing harm, and not because of their ease of being approached from some practical or legal standpoint. And there is always more than one way to look at these issues. Thus the author has strived to be objective in telling the story and in the analysis afterward. As the dredging program manager at EPA, the author observed the points of view of all the relevant players in the dredging crisis and was closely involved in the developments as they occurred.

ACKNOWLEDGMENTS

There are many people who have been involved in the story and events of this book, and that I have worked with and learned from. Although the book was written largely from daily logs kept during my work at EPA, with some research done afterward, the people mentioned below are among those that come most to my mind. Several individuals also were helpful in updating me on events and situations in the Port after I left NYSDEC, where I worked during my last period of involvement with the Port dredging scene. The ones mentioned first will probably be amused to see their names in the same sentence: Monte Greges, dredging section chief of the New York District Corps of Engineers (my first federal supervisor), and Kaydee McGuckin, dredging permit manager at the Region 2 New York State Department of Environmental Conservation. Others who have been helpful in updating me on recent events in the Port (or reminding me of past ones) include Eric (Decon Man) Stern and Charles "Buddy" LoBue of EPA Region 2, Joe Olha of the New York District Corps, and Matt Masters of the PANYNJ. I would also like to thank Paul Dragos, senior Research Scientist, Battelle for the ocean survey photos that he dug up from archives and for his patient explanations during long phone calls of fine points in physical oceanography prior to the 1990 winter surveys. (Paul is not, by the way, the physical oceanographer referred to in the discussion of the big nor'-easter that ravaged the Mud Dump Site.) The biggest factor in doing the book was actually working through the dredging crisis with my colleagues at EPA and in the other agencies. Of these there are too many to mention, but the experience would have been much less bearable without them.

Mario Del Vicario, as my branch chief through this period, had the wisdom to grant me liberal reins in the program (I smile in saying that), while providing needed support and good advice. EPA biologist researcher Norm Rubinstein at the Narragansett laboratory provided much very helpful advice and context, as well as scientific background on many issues after I was first turned loose on boxes of dredging program files and manuals. Rich Pruell, a chemist researcher at Narraganset, has given me valuable insight into chemistry questions that came up more times than I can remember. He also conducted, with Norm, the prominent dioxin bioaccumulation research study discussed in the book. Phil Cook (retired), a researcher at EPA's Duluth laboratory who conducted the Lake Ontario dioxin food chain project shared his theory of bioaccumulation from thin contaminant microlayers on sediment surfaces and other insights from his work. Vic McFarland and Bob Engler, researchers at the Corps' Waterways Experiment Station in Vicksburg, proved to be both knowledgeable scientists and worthy adversaries on some environmental issues. Vic was also largely responsible for much of the work behind theoretical bioaccumulation potential, and Bob, though long since a manager at WES, managed also to keep well abreast of the latest science. I would like to thank these and the other people that I have worked with before, during, and after the main stage of the dredging crisis for helping to make the experience something worth writing about.

I also want to thank Frank Reilly for his critical reading of the manuscript, and for his valuable input during the publication of the book.

ACRONYMS AND ABBREVIATIONS

ALS	American Littoral Society
CA	Coastal Alliance
COA	Clean Ocean Action
CORPS	U.S. Army Corps of Engineers
EDF	Environmental Defense Fund
EPA	U.S. Environmental Protection Agency
ERL-N	U.S. EPA Environmental Research Laboratory, Narragansett (R.I.)
FDA	U.S. Food and Drug Administration
FWS	U.S. Fish and Wildlife Service
HARS	Historic Area Remediation Site
HSV	Hudson Shelf Valley
MDS	Mud Dump Site
MPRSA	Marine Protection, Research, and Sanctuaries Act of 1972
NJDEP	New Jersey Department of Environmental Protection
NMFS	U.S. National Marine Fisheries Service (of NOAA)
NOAA	U.S. National Oceanographic and Atmospheric Administration
NYSDEC	New York State Department of Environmental Conservation
NYSDOH	New York State Department of Health
OMB	Office of Management and Budget
PA DEP	Pennsylvania Department of Environmental Protection (of U.S. Congress)
PANYNJ	Port Authority of New York & New Jersey
PN/E	Port Newark/Elizabeth
RA	Regional administrator (of EPA)
WES	Waterways Experiment Station (of U.S. Army Corps of Engineers)

INTRODUCTION

The Port of New York and New Jersey is one of the busiest cargo ports in the United States. Over 78 billion tons of cargo, valued in excess of $100 billion, passed through the Port in 2003, and these numbers are projected to continue growing every year. It is the largest container port on the East Coast. More cargo now moves by truck-size containers than any other way, and this trend is also continuing. The total capital value of the Port has been estimated at more than $29 billion, and it is directly or indirectly responsible for approximately 167,000 jobs, split about evenly between New York and New Jersey residents. In 1997, New York exported $54 billion worth of goods to two hundred countries through the Port, and New Jersey exported $22 billion, making them the third- and ninth-largest exporting states in the country. The Port is the largest petroleum-handling port in the country. It is, in short, a very vital component in the economy of both states.

The Port is also, from an environmental standpoint, part of the New York/New Jersey Harbor Estuary ecosystem (figure 1). An estuary is a partially enclosed body of water with a connection to the ocean where fresh water from one or more rivers mixes with salt water. Estuaries are widely considered to serve important functions as reproduction and nursery areas for marine life and, thereby, substantially affect the larger surrounding regions. The ocean just offshore of the Port estuary is an important recreational and commercial fishing area for the coastal residents and for visitors. (A terminology note here: "New York/New Jersey Harbor" or "the Harbor" are the terms usually used in association with the Port's environmental aspects, as in "Harbor Estuary," so in keeping with this practice the book will use those terms as appropriate, but for all intents and purposes the "Port" and the "Harbor" encompass the same

FIG 1. The Port of New York and New Jersey. (Courtesy the New District Corps of Engineers, Public Affairs Office.)

general geographic area.) When it comes to numbers of coastal residents, the New York/New Jersey metropolitan area has among the highest concentrations in the country. So the environmental aspects of the Port are often in direct conflict with its industrial and economic needs. These conflicting aspects, worked by the proponents of the various economic and environmental interests, and the congested land base of the metropolitan area, brought about the dredging crisis in the Port.

An essential fact in this story is that, to maintain navigational depths for modern shipping, channels and berths have to be dredged, and the dredged material has to go somewhere. In most industrial ports, bottom sediments are contaminated to various degrees by pollution from industry and the general population within the port area or upstream (or upwind) of it. Pollution comes into the Port from rivers containing agricultural pesticides and industrial contaminants, from outfalls that discharge industrial and other contamination from immediately surrounding areas, and from air pollution originating in both upwind regions and the Port vicinity that settles on adjacent land and the large surface water areas. Many of these pollutants attach themselves to suspended particles in the water and can sink to the bottom and become sediments (and future dredged material). In estuaries such as the Port, sedimentation from rivers is high in some areas because of the drop in current speeds as the rivers widen into the estuary. Sedimentation rates vary widely within the Port because of the complexities of the currents and tidal flows, but in many places dredging has to be done every several years or so to maintain depths adequate for navigation.

In the Port of New York and New Jersey, as in many others in the United States and worldwide, almost all dredged material has been dumped in the nearby ocean. As national environmental laws for ocean dumping were tightened in the early 1990s, however, and then implemented for the Port, alternatives to the ocean had to be found. Since the Port's surrounding land area is congested and land disposal alternatives are scarce, it would be expected that finding disposal options for dredged material would be difficult. Adding to this difficulty was the fact that it takes time to develop land-based disposal options, and there had been very little done to develop alternatives in the period leading up to the 1990s. When dredging was virtually shut down for a few years in the early to mid-1990s, the Port's economic viability was seriously threatened. It was obvious that if shipping was going to be restricted from moving into and out of the Port because of the siltation of channels and berths, the Port's business would soon wither. This would have devastating economic effects for the entire New York metropolitan area. It could not be allowed to happen.

Even the threat of a shutdown can actually accelerate a port's decline, because ports generally are very competitive. A loss of business for New York could be a boon for Baltimore or Norfolk or Halifax. Add to this that modern tankers and container ships are steadily increasing in size and draft and therefore need deeper ports, and you have the potential for a chain-reaction crisis. Shipping businesses are true examples of long-range thinking. They have to be, because the nature of their business depends on it. Shipping deals are struck for years in advance, and there is a lot of cooperation and interdependency, as well as competition, among shippers. If shippers begin to believe that the future of a particular port is questionable due to projected difficulties in maintaining or deepening channels and berths, they will be hard-pressed to stay loyal to it. They will seriously consider moving vital facilities to a more promising port, one with fewer operating restrictions for the near and long term. If such a shift starts to happen, it can snowball quickly among shippers in a port, causing a drastic reduction in the port's economic operational level. This could result in its no longer being a "hub" port.

The larger ports are hubs in the sense that they are where international import/export shipping and land-based truck or rail distribution and receipt of goods takes place. They can also act as hubs for distributing cargo to smaller ports. Smaller ports, or those with disadvantages such as having shallow water, inadequate port facilities, or poor land transportation infrastructure, do not get the big ships or big business. The loss of a port's status as a hub port could have grave economic consequences for the region that it services, which, for the Port of New York and New Jersey, includes the New York metropolitan area and well beyond it.

1

DREDGING AND OCEAN DUMPING

THE WAY WE WERE

It was a beautiful June day in 1989 on the trendy Annapolis waterfront, and the meeting room at EPA's Chesapeake Bay Program Office was about three-quarters full. The EPA headquarters' Office of Marine and Estuarine Protection was holding a preliminary get-together with its ocean dumping coordinators from EPA's coastal regions around the country. At the big annual meeting to be held the next month, EPA headquarters staff and the regional coordinators would be joining their counterparts from the Corps. In the national ocean dumping program, "counterparts" is an apt term in describing the prevailing relationship between the two organizations. This is because these two federal agencies are bound by law to jointly oversee the national dredging and ocean dumping program, while also being driven by diametrically opposing missions.

The Corps of Engineers, an agency that was born around the time of the Civil War, has evolved from essentially a road- and bridge-building subsidiary of the U.S. Army to one with the broad-ranging missions it carries out today. The Corps' primary mission, however, is still to build things as part of its Civil Works Program in areas of its jurisdiction, which encompasses essentially all U.S. coastal and inland waterways. The agency also administers a national permitting program that regulates any construction proposed by other parties in its jurisdiction, which also includes all U.S. wetlands. One of its major missions is to keep federal navigation channels clear by dredging them to depths adequate for shipping. So, not only does the Corps have an interest in promoting and conducting dredging operations itself as part of its federal navigation program; it also regulates dredging and related applications by private entities and other government and nongovernment organizations through its regulatory permit program.

In conducting these missions, the Corps is required to take into consideration a number of factors that affect the general public welfare, mainly economic in nature but also including environmental issues. These last have been fairly recently added to the Corps' scope of factors to consider in its decisions, via several dictums from Corps headquarters. These dictums, mainly in the form called regulatory guidance letters, officially require field offices to more fully consider the environmental effects of projects they want to carry out, or that they would permit applicants to conduct. This historically new twist in official Corps planning and thinking has sometimes been slow in gaining full acceptance by the "old guard" in the Districts.

The EPA was formed in 1970, mainly to address the increasingly nasty water and air pollution problems that were threatening the environment and endangering people's health. Those were the days when the Cuyahoga River, for example, caught on fire because of its toxic mix of industrial effluents. It is still EPA's primary mission to protect the country's environmental resources and human health. EPA plays a large role in government decision-making on dredging projects, especially those that propose to dump the dredged material in the ocean. The Corps must abide by regulations that Congress stipulated to require the two agencies to share authority in this area, with built-in checks and balances. The regulations give EPA the authority to veto proposals to dump materials, including dredged materials, into the ocean. This puts EPA directly in the Corps' face, so to speak, since it forces the Corps to share one of its primary mission responsibilities with the environmental agency. The two agencies also jointly manage the Clean Water Act (1972) wetlands program, but in a different regulatory relationship.

The Ocean Dumping Regulations (EPA 1977) were written to protect the ocean from "unacceptable adverse impacts" that might result from any regulated disposal activities that would take place in it. The Corps had a consultative role in setting the regulations and retained a great deal of authority in the area of dredged material dumping. EPA was empowered to administer and enforce the regulations, meaning that EPA can veto a Corps (or any other) project if it believes that the material to be dumped or the methods for doing it will not meet the regulatory requirements. There are provisions that acknowledge the Corps' important role in the world of dredging. These give the Corps the ability to appeal a veto, and they specify what EPA must do to override the appeal as the case elevates to higher agency levels. By the regulations EPA would normally have the last word at the end of this process, if it felt it had a strong enough case. However, agency managers generally tend to avoid elevating cases to higher authority if it is at all possible to resolve them at their level, since elevated (headquarters) decisions that go against initial (regional) findings can be detrimental to a regional manager's career path. Therefore, there have been very few cases that have been elevated to the respective agencies' headquarters in the national dredging/ocean disposal program.

The Corps has the responsibility of dredging navigational waterways, and in many cases, as was noted, the most practical place to dispose the dredged material has been the ocean. However, Congress also recognized that this could have negative effects on the marine environment, especially if the material was contaminated. In many industrial harbors, water pollution that has existed over long time periods has in fact contaminated the bottom sediments. While in many areas around the country implementation of the 1972 Clean Water Act has resulted in cleaner waters, sometimes remarkably so, sediment contamination remains a lingering and difficult problem.

EPA was therefore given the role of riding herd on the Corps to try to ensure that detrimental effects to the environment do not occur from ocean disposal of dredged material. The U.S. coastal oceans, though no longer pristine, are relatively much cleaner than most industrial harbors, and the public has become increasingly determined to ensure their protection from human abuses. A major law and the resulting regulations were therefore written to try to provide a balance between the two agencies and their divergent missions. However, the law, the Marine Protection, Research, and Sanctuaries Act (MPRSA) of 1972, and the duly promulgated Ocean Dumping Regulations, are in many instances, unfortunately, not shining examples of clarity. Large portions are therefore open for interpretation by both agencies when disagreements develop, and those interpretations usually go according to the respective agency's missions and goals.

The regulations were developed from the law, and a national testing program was developed from both to evaluate whether materials are suitable for ocean disposal. The testing and evaluation requirements are quite scientific and technical, but as in many areas of science, there are not always clear answers to every issue. Since there remain many issues that are open to judgment in the evaluation of test results, and even sometimes in how to conduct the tests, points of contention have developed on a regular basis. There also exist scientifically based requirements for designating and managing an ocean disposal site, duties that are also shared between the two agencies. Therefore, periodic meetings are held between the two agencies to try to iron out the many differences of opinion between them. The get-together in Annapolis was in preparation for one such meeting.

Representatives from EPA coastal regions based in Boston, New York, Philadelphia, Atlanta, Dallas, San Francisco, and Seattle, and of course from headquarters in D.C., were there. As usual, a good number of policy and technical topics were discussed. Most of these involved the planned issuance of a newly revised testing manual for dredged material proposed for ocean disposal. Although many technical and scientific issues had still to be worked out, it was clear to everyone that the revisions being discussed would result in making it more difficult for dredged material to pass ocean disposal testing. How much more difficult it would be depended on how contaminated a particular port's sediments were, what level of testing was currently being done in the region, and, to some degree, how the national guidelines would be specifically implemented in the different regions. Apparently in keeping with many other recent national trends, the two West Coast regions were leading the way and starting to implement the more scientifically advanced procedures that were planned for the new testing manual. It was not always clear, however, how the results of the new test procedures were being evaluated by these two regions. This is important because how tests are evaluated can be just as telling as whether they are being used in the first place.

The reader may find it curious that there would be such differences in the way a national agency carries out its program in different parts of the country. The regional EPA offices do have a large degree of freedom, within bounds, to carry out the national program. The main reasons cited for these differences are to take into consideration the differing environmental settings and local regulatory (state and local) backdrops. For example, the differing climates and marine environments around the country may call for the use of different regional test organisms, which could require using different test conditions. (Test conditions for all the bioassay tests are very specific in the regional manuals, down to an allowable range for variables such as water temperature and pH.) Also, some regions are located where there are relevant existing state or local environmental regulatory programs that could benefit by being adhered to as much as possible by federal agency programs. And one thing that is shared by EPA and its "sister agency," the Corps of Engineers, is this phenomenon of fairly wide-ranging differences in regional implementation policies.

There existed also during the late 1980s, besides a difference in missions, a difference in what was at stake for the two agencies regarding the dredging program. For the Corps, navigational dredging is a major responsibility and a major source of congressional budget allocations. Although the Corps' heart normally sides with the applicants in its regulatory permit program and the Corps will fight for them, another part of its anatomy, so to speak, gets restricted when a major source of its annual funding is threatened. Also, it is clear that most members of Congress are sympathetic to the importance of the Corps' mission. It does not take many letters from wealthy boaters or, more crucially, from large shipping companies complaining about navigational and related economic restrictions to get a congressman motivated. EPA, on the other hand, did not in the late 1980s have a large stake in the dredging program, neither fundingwise nor as an environmental priority. It began to assert itself more as scientific information on the problem of sediment contamination came to light, but the Corps was still by far the dominant player at this time.

There had been earlier squabbles with EPA about regulations and joint guidance development, as well as negotiations with EPA and other federal and state environmental agencies over particular projects through the years. Throughout all of these, the Corps of Engineers could feel fairly confident that it would prevail in matters that could be perceived as threats to its program. Part of the reason for this was the commonly held belief at the time that contamination in dredged material was not really a serious problem. The prevailing philosophy at many of the state and federal environmental resource agencies (including the U.S. Fish and Wildlife Service and the U.S. National Marine Fisheries Service) was that there were other environmental matters more pressing and serious than dredged material. (In fact, an EPA manager had told a new staff scientist assigned in 1988 to the dredging program that dredging was the least of his worries. That would change beyond the manager's wildest imagination.) The underlying and broader issue that encompasses dredged material, contaminated sediments, soon began to receive intense and continuing attention from EPA. Other resource agencies, especially the Fish and Wildlife Service, would also recognize the problem as a major national environmental issue. Prior to the mid- to late 1980s, however, dredged material was still a rather specialized realm mainly controlled by the Corps, with some inhibited interaction by EPA and lesser interaction by the other resource agencies.

Another reason that the Corps was not overly concerned about potential problems raised by environmental resource agencies was that, in large part, the Corps had science on its side. That is, it had the Waterways Experiment Station in Vicksburg, Mississippi. WES was a state-of-the-art engineering and science laboratory complex responsible for conducting research, engineering, and development for many of the Corps projects and programs. (It has had its name changed to the Engineering Research and Development Center, but for period authenticity and to avoid confusion it will continue to be referred to here as WES.) The Corps has a couple of other engineering centers around the country, but most of the coastal hydrology, environmental, information technology, geotechnical, and structures work is done at WES.

Scientists at WES had conducted scientific investigations on all aspects of dredging and dredged material disposal for years and had published numerous articles in both their own in-house (gray literature) publications and in peer-reviewed science journals. During that time at EPA, work was being done by a few research scientists on some aspects of dredged material ocean disposal and on contaminated sediments, but it was on a much smaller scale and enjoyed considerably less funding. It was not yet a hot environmental issue. But it was a very important issue to the Corps of Engineers. It was very much in the Corps' interest to remain in the lead scientifically, both for the knowledge that it provided to respond credibly when concerns were raised by resource agencies or the public and for the prestige it afforded the agency to help it more easily dispense with any such troublesome issues.

Last but not least, the Corps knew that if push came to shove, it would be easier for the Corps to claim that ocean disposal was in the public interest than it would be for a resource agency to demonstrate an overriding environmental harm. The scientific information available at the time did not seem to support the contention that disposal of dredged material, other than the relatively rare "sediments from Hell" (colloquialism for the most contaminated sediments), posed any substantial risk of environmental harm. The most current science regarding the effects of dredged material dumped into the ocean or even the effects of dredging itself was generated mainly by the Corps from studies conducted at WES. The tests that were in place in the existing, original version of the national testing manual, the "Green Book," were not hard to pass. The Green Book had been established and used since 1977 as the national testing guide for dredged material proposed for ocean disposal (it originally was issued with a greenish cover). There was very little hard science available that could demonstrate "unacceptable adverse impacts," as is required by the regulations to keep materials from being dredged and disposed of in the ocean.

So the Corps at that time was fairly confident that it could contend with questions raised about its own projects or about important permit decisions, such as large PANYNJ dredging and disposal projects. This situation was not to last much longer.

Although the testing manual was the main topic of the Annapolis meeting in 1989, there were undercurrents of two other important issues running just below the surface. One of these was the discovery in dredged sediments of a notoriously dangerous chemical, one that the public had come to know and fear. Dioxin, a by-product formed in minute quantities during certain chemical manufacturing processes, was being found in sediments and marine life in the waters of New York and New Jersey. (Dioxin is also produced in less concentrated and usually less harmful forms during some natural events such as forest fires and volcanic events.) The public had first heard about dioxin years prior to this from a series of articles written about it and from the health problems that Vietnam veterans were claiming had been caused by Agent Orange, a defoliant used during the Vietnam War. Veterans of that war, especially those who had applied the defoliant in Vietnam, were experiencing a range of maladies, from persistent skin rashes to cancer.

There was scientific evidence that put the blame for these ills on dioxin, a by-product of Agent Orange production. The most important hazard of dioxin known at the time was that it was a carcinogen. It was also found to be the most powerful of all known acutely toxic (immediately acting) poisons, natural or synthetic. These kinds of facts tend to remain in the public memory. The finding of dioxin in sediments in Newark Bay, a major shipping area within the Port complex, ensured that regulators in the dredging program would have their hands full before too long. This one issue, perhaps more than anything else, would shape the events of the next several years in the Port of New York and New Jersey.

Another underlying issue that was on the ocean dumping coordinators' minds was EPA's loss of a lawsuit several years prior, brought by the National Wildlife Federation, that had required EPA to rewrite entire parts of the Ocean Dumping Regulations. The lawsuit complaint was that dredged material was treated differently in the regulations from other material proposed for ocean dumping. (The changes that were made in response to this suit actually created problems for EPA in a later lawsuit brought by other environmental groups, but more on this later.) Government agencies do not like to lose lawsuits any more than anyone else does. This court case was still fresh in the agency's institutional memory, and it probably helped to create a climate in headquarters of heightened sensitivity to challenges by environmental groups. It also put EPA squarely in the middle of what would become a continuing relationship: the environmental groups on EPA's left, pushing for stricter environmental controls to protect the ocean, and the Corps on EPA's right, either resisting any change whatsoever or proposing alternate and more relaxed changes that would maintain a regulatory climate more suited for the continued agreeable performance of its mission.

Meanwhile, EPA's scientists and others were reaching consensus that the existing testing methods were seriously inadequate and out of date. One type of bioassay test that was required, for example, measured the percentage of dead organisms in a group exposed to a sample of the dredge sediments proposed for ocean dumping. The type of clam that was commonly being used as a test species was so hardy that, the joke was, you could hit one with a hammer and it would not be fazed. Other marine species were also used in other tests, but this particular organism did not seem like a good example of a sensitive species, as required by the regulations. The environmental groups were, for their part, beginning to demand an end to ocean disposal of dredged material. They were referring, really, to contaminated material, but most New York Harbor dredged material was contaminated to some degree, and many in the agencies believed that the environmentalists actually wanted to stop all disposal of dredged material in the ocean and would hold this view no matter how much more restrictive the testing became. This was the enviro-political situation that the ocean dumping coordinators found themselves in, and many of them figured that some major convulsions were in the wind.

By 1989, the dredged material ocean dumping program in the New York area was only beginning to get the attention of the environmental community. There were other ocean dumping activities still going on that had to be attended to first. The Industrial Waste Site, located 106 miles offshore, and the Acid Waste Site, located 17 miles offshore, were being prepared to be de-designated, or officially shut down. For virtually any kind of proposed ocean disposal to occur, EPA first has to designate a site, which usually requires fairly extensive investigations, including oceanographic studies. Each designated ocean disposal site is then listed in the Ocean Dumping Regulations. If no further use of a designated site is foreseen, or if other reasons dictate its closure, it has to then go through a formal de-designation process. For some kinds of minor, one-time disposals, such as burial at sea, the process is less formal.

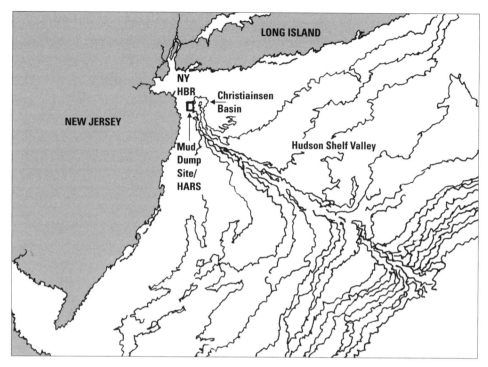

FIG 2. The New York Bight coastal shelf and shelf break. This figure shows enhanced depth contours taken from a NOAA chart of the N.Y. Bight. The selected and enhanced contours show areas of equal water depth throughout the continental shelf to the shelf break. The more closely spaced contours running diagonally down toward the right lower corner indicate the Hudson Shelf Valley, and the wider area of close contours running in the opposite direction on the lower right side depict the shelf break. (Adapted by the author from NOAA photo archives)

Industrial wastes had been dumped since 1961 at the 106-mile site, but with increasing pressure from environmental groups and EPA's increasingly strict monitoring and reporting requirements, by 1988 only two companies still used the site, Dupont and Allied Signal. These companies disposed of various industrial wastes generated from chemical processes, including acidic wastes that contained heavy metals. In that year, Dupont withdrew its permit renewal application, and Allied followed the next year, opening the way for de-designation. In this time period the 17-mile site, which had been designated for disposal exclusively of acid industrial wastes, was also de-designated.

Meanwhile, burning of Port driftwood and old piers and ocean disposal of the ashes and residues was allowed under the regulations, starting in 1968 at a designated site seventeen miles offshore of Point Pleasant, N.J. A wide range of interests including fishermen, boaters, and beachgoers had opposed this practice since around 1984. This was when Clean Ocean Action, a New Jersey coast environmental group that would eventually lead the effort against ocean disposal of dredged material, started a campaign against it. The Corps of Engineers terminated the last ocean wood-burning contract in 1991.

Clean Ocean Action started as an offshoot of the coastal environmental group American Littoral Society. Headed by Cindy Zipf, it was dedicated to reducing or eliminating ocean disposal in the ocean waters off New Jersey. COA and other environmental groups were also trying to close down the dumping of sewage sludge, which had been going on in the New York Bight since 1924. This dumping took place at a site twelve miles outside New York Harbor until the area became so polluted that in 1987 the dumping was moved to a site 106 miles offshore. The magic number of 106 miles is the approximate distance offshore where the relatively shallow continental shelf breaks sharply downward to the abyssal plain. There, much deeper waters would theoretically more quickly dilute any dumped materials (figure 2).

A number of New York and New Jersey municipal authorities were dumping at these sites; New Jersey stopped in March 1991, and the last New York authority followed suit in June 1992. Now that these sewage sludge and other ocean dumping activities were shut down, Clean Ocean Action and other groups were free to start homing in on dredged material. (Incidentally, to dispel some apparently common misconceptions, municipal solid waste that typically goes to landfills or incinerators has not been allowed to be disposed of in the ocean since 1934.)

2

CONTAMINATION

IT'S DIOXIN, STUPID

The Diamond Alkali Company, later known as the Diamond Shamrock Chemicals Company, and now known as the Occidental Chemical Corporation, owned and operated a pesticide manufacturing plant in Newark, New Jersey, from 1951 to 1969. In 1983, the New Jersey Department of Environmental Protection and EPA found dioxin on the site and in the adjoining Passaic River, a tributary of Newark Bay. Newark Bay is connected to New York Harbor proper via the Kill Van Kull (to the upper Harbor) and the Arthur Kill (to the lower Harbor). (These two kills are what makes Staten Island an island. The word "kill" comes from the Old Dutch term meaning "stream.") It was believed that much of the dioxin sediment contamination in Newark Bay and beyond, to other parts of the Port, came from this site, spreading with the river flow and tidal currents. Diamond Shamrock was one of the seven chemical companies that in 1984 settled out of court a class-action suit brought by victims and families of those exposed to Agent Orange in Vietnam. The victims and families won a collective $180 million (the legal fees were reportedly $100 million).

The NJDEP discovered in the mid-1980s that crabs and fish in parts of Newark Bay had alarming levels of dioxin in them. In August 1986 the New York District Corps assembled the Federal Interagency Dioxin Steering Committee to try to address the public's concerns and the scientific questions on how to deal with dioxin in dredged material. The committee was chaired by the New York District Corps and represented three other agencies at its inception: EPA, the U.S. Fish and Wildlife Service, and the U.S. National Marine Fisheries Service, with the states of New York and New Jersey environmental agencies included in an advisory capacity. Eventually, the Corps included the New York State Department of Environmental Conservation and the New

Jersey Department of Environmental Protection as full members. (Even though these were not federal agencies, the Corps kept "Federal" in the name of the committee, perhaps as a gesture that it meant for the feds—meaning the Corps—to be in charge.) It was, after all, the NJDEP's finding of high levels of dioxin in Newark Bay biota that had raised the red flag of dioxin over the regional dredging program. The relatively quiet control that the Corps held over the program for years would soon come to an end. EPA and eventually the states would all feel the heat from the public's alarm over these findings, voiced mainly by environmental groups, local newspapers, and some New Jersey congressmen. It was the specter of dioxin in New Jersey fish and clams, in an area that took its seafood seriously, that would be the initial driving force for some serious questioning of the Corps' and EPA's regional ocean dumping program.

A major stakeholder in the business of the Port is of course the PANYNJ. Newark Bay, the area that was believed to have received the majority of the dioxin contamination downstream from the Passaic River site, is the home of the Port Authority's two largest shipping terminals, Port Newark and Elizabeth Terminal. The PANYNJ was therefore very much concerned and would soon become even more involved than it had been in the management of the Port's dredging program. It was not a member of the Federal Interagency Dioxin Steering Committee, since it is not federal and is only a quasi-government agency.

The New York District Corps had, of course, always worked closely with the PANYNJ on particular projects and on strategic planning regarding dredging the several large shipping terminals and facilities around the Port. The Corps also coordinated with other, smaller applicants for federal dredging permits, to clarify the testing requirements for ocean dumping or to resolve questions about the permit process. The PANYNJ was, however, a major player within the Corps permit program, and it stands to reason that, because of the large amount of dredging done by both organizations, there was much more coordination between them than between other typical permit applicants and the Corps.

The PANYNJ had started its own committee, which it named Save Our Port. Invited to meetings of this group were the Corps, representatives of shipping companies that were tenants of the Port Authority's terminals, the Coast Guard, assorted other shipping or dredging interests, and, for a while, EPA. These meetings were mainly about reviewing past dredging projects and strategizing for future projects that were needed to maintain and improve the Port. Although EPA was politely invited, the reason seemed to be mainly for the purpose of instilling in the minds of EPA regulators the importance of the Port dredging projects, and not particularly to gather valuable insight from EPA on possible environmental issues. This was made rather clear in a meeting that was attended by an EPA representative (the author) in 1989. Leading up to this time, there had been much discussion and many draft revisions circulated among the federal ocean disposal coordinators indicating that the revised national testing manual could result in drastic changes in what would be acceptable material for ocean disposal. In the 1989 roundtable session of the meeting, EPA noted that progress was being made by the Corps and EPA in drafting a revised national testing manual and that it might be prudent to begin preparing for the possibility of dredging projects failing more stringent testing by looking more seriously at nonocean disposal alternatives. The reaction by the group was at first a stony silence. A comment finally came from a PANYNJ representative that whatever was going to be in the revised national testing manual was after all only guidance, not regulation, and it was the regional Corps' and EPA's choice whether to implement any of it in their regional testing manual.

It is true that the national testing manual, the Green Book, is in fact guidance and does not have the force of law. The Green Book, however, was being faithfully and literally implemented in the regional testing manual, as was the case generally across the country. In fact both the Corps and EPA held the Green Book to about the level of regard that the Pope holds the Bible, and they did treat it as de facto law. This was because Corps as well as EPA scientists were responsible for developing it, and it had served them well thus far. It contained testing methods that used bioassays, which had become considered the best methods for conducting sediment environmental evaluations, and so gave the testing program an aura of not only environmental responsibility but even, for a time, progressiveness. Therefore, for anyone involved in Port dredging issues to believe that the first and only Green Book revision in twelve years could be of no real consequence seemed (to the author) to portray either extreme naïveté or serious denial. All the recent indications, including the growing turmoil in the environmental community and the accord among the agencies' scientists that the revisions were necessary, pointed to a high probability that the revisions would indeed be implemented in a revised regional testing manual.

Despite those apparent driving forces, it is not difficult to understand why there would have been an aversion to seriously considering nonocean disposal alternatives. Nonocean alternatives had already been evaluated: an in-depth study of alternative disposal methods for the Port's dredged material had been funded by the New York District Corps and carried out by a respected environmental consultant. Everything from disposal islands to filling in borrow pits (left from prior sand-mining operations) to shoreside containment facilities to various land disposal methods was identified and carefully evaluated. The Corps had provided the consultant with evaluation criteria for how to assess such options. These included things like the minimum size for a parcel that could be considered for land disposal options and the presence of any wetlands on a site. The final report described the probable impediments from both regulatory and public acceptance standpoints, and included the projected costs of various scenarios in each of the possible options. The conclusions of the report indicated that there were problems with all the options. They were all deemed to be either too difficult or impossible to implement from a regulatory or public acceptance basis, or their costs were prohibitive compared with the cost of ocean disposal.

Many of the candidate sites were disqualified by the size requirement or the requirement that no wetlands be located on the site. A concern that EPA staff had with the report, however, was that requiring a large site size obviated the possibility of identifying a number of smaller, localized sites instead of the large regional facility or facilities that the Corps favored. The wetlands restriction seemed appropriate on its face, but it could have resulted in disqualification of sites that had only a small amount of wetlands. This might not have constituted an actual regulatory impediment if the wetlands portion was limited enough that measures could be included for its protection or exclusion from being affected by the site's use. Region 2 EPA sent a letter (prepared by the author) to the New York District Corps outlining these concerns.

Disposal costs were the main impediment for most nonocean alternatives, regardless of the other factors. The long-used regional ocean disposal site for dredged material, a 2.2-square-mile area called the Mud Dump Site, was located about six miles off the northern New Jersey coast. The cost for disposing of a cubic yard of dredged material at the Mud Dump was approximately four dollars. No other option had a cost that would be anywhere near this low. Even disposal in Harbor borrow pits, much closer to the dredging sites than the Mud Dump, was calculated to cost more than ocean disposal. This was because more carefully controlled operations and monitoring, as well as capping with clean material, would most likely have to be included. When it was confirmed that costs for the alternatives ranged from anywhere around three times to more than ten times the cost of ocean disposal, the enthusiasm in the search for alternatives cooled somewhat.

The revised testing issues included the evaluation of dioxin in dredged materials. An important aspect of the testing issues in general, but first and most critically raised for dioxin, was the subject of detection limits. Environmental laboratories had been increasingly improving their ability to detect contaminants in water, sediment, or animal tissue at very minute levels. One of the revised testing requirements was that labs had to achieve the lower detection limits that were currently becoming routine in the better laboratories around the country. The Corps and the dredging community did not greet this requirement with open arms. The ability to detect contaminants at lower levels meant that more contaminants would be found in proposed dredged materials.

Many contaminants are now found almost universally in all media—air, water, soil, animal tissue (including humans)—when the detection level is low enough. The question from an environmental and regulatory standpoint is, what levels are cause for concern? Is any level that can be detected a legitimate concern, or some other level above the smallest detectable amount? Some scientists who had studied dioxin had said they did not think there was any safe level for it as a carcinogen. In toxicology the term is that the response is nonlinear; no matter how minute the quantity, as long as some amount is there, it has the potential of causing an effect, such as cancer. Of course there were, and are, a few scientists that look at the same data and do not take as restrictive a view.

EPA was in fact soon to embark on a huge national reassessment regarding the appropriate regulatory standards for dioxin. This would include a review of all the previous toxicological data as well as new data that had more recently been developed, and it would be done by agency as well as outside experts with full public review and opportunity to comment at several key stages. Out of it would come technical recommendations for changing or not changing the EPA dioxin cancer risk levels and other risk criteria that the agency was using to determine hazardous or potentially harmful conditions. Other EPA programs besides dredging/ocean disposal were using these risk levels for permitting and evaluations, including other water programs that regulate discharges from industrial and sewage outfalls, as well as land-based remediation, hazardous materials management, and air programs. The controversy over the issue of nonlinearity has continued, as industry groups have protested and brought suit against EPA in subsequent dioxin assessments and proposed rule-making.

At this point in time however, in 1989, there was one toxicological study that had been done on rats (Kociba and Schwetz 1982) available, from which cancer risks were extrapolated to humans. There was not a lot of other scientific information on which to base decisions regarding low levels of contamination. For dioxin, the detection level that could be routinely achieved by labs had fallen significantly during the 1980s, from about 50 parts per trillion to about 1 part per trillion. That's one part dioxin to one-with-twelve-zeros-after it parts of whatever the dioxin is in—water, sediment, or animal tissue. Someone at the PANYNJ, which had its main offices at the World Trade Center (WTC), had done some calculations and developed a statistic that was trotted out on good occasion. One in a trillion equated approximately to a golf ball in the space of one of the Trade Center towers. This analogy might have made an impression in some circles but was met with no more than mild amusement by most agency regulators that it was used on. (Personal note: This passage was originally written before September 11, 2001, and for someone who witnessed the second tower go down, it was not a casual decision to leave it in. It is, however, a good reflection of the extent to which the WTC occupied the mind-set of New Yorkers even pre-9/11.) It is true that these concentrations are almost inconceivably small, but scientists had already known that EPA water quality criteria for dioxin were set at parts per trillion levels. Nevertheless, measurement to these levels could only be estimated by using extensive sample preparations and the most sensitive analytical methods.

The Federal Interagency Dioxin Steering Committee was formed to try to reach consensus between the Corps and the resource agencies about what level of dioxin would result in a project's failing the ocean disposal criteria. The Corps initially floated a straw man proposal to the group that appeared to follow the general guidelines in the Green Book. The guidelines were based on measuring how much of a contaminant accumulates in the bodies of marine animals exposed to the test material. A sediment can be contaminated, but the contaminants may not be readily accumulated in marine organisms if, for example, they are tightly bound to sediment particles and are not readily available for organisms to take in. Sediments vary in their ability to bind contaminants, based on various physical and chemical properties. So instead of directly analyzing sediments for their level of contamination, regulators have analyzed the tissues of test animals to measure how much of a contaminant was bioavailable and was accumulated. This uptake process is termed bio-accumulation, and the routine tests for measuring this are conventionally now run for a one-month period. The bioaccumulation test in the old Green Book required only a ten-day test period.

The Corps and EPA are careful to describe bioaccumulation as a "phenom-enon," not an effect. Unlike mortality, for example, bioaccumulation is not an effect. Contaminants can affect organisms in variable ways, according to the concentrations in their tissues and according to chemical variations among dif-ferent kinds of organisms. Other tests that are done for dredged material are called toxicity tests, and they always report results by measurement of an ob-served effect, whether it be mortality, low or abnormal growth, low reproduc-tion, or abnormal behavior. These toxicity tests are usually done with sensitive organisms that die or are otherwise adversely affected by appreciable pollution levels, and the severity of the effects determines whether the material passes or fails the test. In bioaccumulation tests, there is no effect measured—only the amount of contaminant in the body is analyzed, and from that informa-tion a prediction of the potential effects in the environment has to be made through various methods. The animals that are used for bioaccumulation tests have to be hardy enough to live in contaminated sediments for a long enough period of time to accumulate contaminants. Remember the clam that was mentioned in the previous chapter that would not be fazed by a hammer? Under the existing procedures in New York (and allowed in the old Green Book), it was being used as a bioaccumulation and a toxicity organism.

There were (and are at the time of this writing) three regional categories of dredged material, which the agencies had named Category 1 (for the cleanest), Category 2 (dredged materials that had to be capped with cleaner material to be allowable for ocean disposal), and Category 3 (materials that could not go into the ocean because they had failed a toxicity test). Categories 1 and 2 were distinguished by the amount of bioaccumulation of contaminants in test or-ganism tissues. (A requirement for both of these categories was that the ma-terial also pass the toxicity tests.) Category 2 was for material that had the po-tential to bioaccumulate to potentially deleterious levels but was nevertheless deemed to be acceptable for ocean disposal with capping. The justification for the capping requirement that had historically been cited was that it would provide an added level of protection from any uncertainty regarding the ma-terial's potential to cause unacceptable levels of bioaccumulation.

The Corps' proposal to the steering committee on how to manage dioxin-contaminated sediments included threshold numbers for the concentration of dioxin in tissues of test organisms. The numbers were 25 parts per trillion (pptr) to qualify as Category 2 and 50 pptr to be classified as Category 3. The practice at the time was that Category 2 material could be disposed of in the ocean and covered with other dredged material soon afterward. This practice was referred to as "de facto capping" and was a subject of as much controversy as capping proper (which requires specific identification of a "clean" material for a cap) has continued to be. Some of the questions raised about capping are, what is "clean" material, how soon must the project materials be covered, to what depth of cap, and, most importantly, how effective is the process? The effectiveness of capping with even clean sand in areas such as the open ocean has been a contentious issue. The Corps' numbers were based on "safe" levels in fish that the U.S. Food and Drug Administration had once suggested in a letter to a state health agency.

These FDA numbers turned into a major issue in the technical and legal arguments over dioxin in dredged material, so a closer look at them is warranted. First, the FDA did not have the level of confidence about these dioxin numbers to include them formally among its "action levels", which were established for a number of other contaminants. (This is still the case at the time of this writing.) Action levels have been established and issued by FDA for a number of contaminants. These are considered by the FDA to be "safe" levels for these contaminants in foods for human consumption. Second, the way FDA derives its action levels differs from the way EPA and most state health agencies derive their criteria: FDA uses somewhat more liberal factors in its calculations. So, the environmental resource agencies did not receive the Corps' straw man with much enthusiasm. A Fish and Wildlife Service scientist who had done work in this area of toxicology was the strongest voice against the Corps proposal. EPA also did not agree with the straw man numbers. It believed they were too lenient according to the science that was developing at that time, and so argument continued between the resource agencies and the Corps through letters and in meetings. Disagreement reigned for over two years, with continuing discussions and the development of several variations of the original Corps proposal, which were put forth by the Corps as well as by other members of the steering committee.

At about this time, when achievable detection limits were heading south toward 1 pptr, other tissue-based dioxin criteria were being developed by various state health departments. The steering committee took note of these. The

New York State Department of Health had recently issued a criterion of 10 pptr in fish consumed by humans. Several other states had issued criteria that were in the range of the New York number. As the EPA representative on the steering committee, the author discussed the derivation methods with the New York State Department of Health (NYSDOH) scientists and developed a new proposed dioxin management plan that included the value of 10 pptr. The plan used the NYSDOH number in a risk-based approach that took into account new scientific information on dioxin bioaccumulation rates over time in the test organisms as well as the probable effectiveness of capping.

The new uptake information on dioxin was from a recently completed long-term exposure study done by the EPA Environmental Research Laboratory in Narragansett, Rhode Island (Pruell et al. 1993). Capping effectiveness was estimated from studies that had been done by the Corps' Waterways Experiment Station laboratory. The proposed plan included the requirement that dredged materials causing any accumulation in test organisms (at a detection level of 1 pptr) would require capping (Category 2), and that materials causing any accumulation greater than 10 pptr would not be allowed for ocean disposal (Category 3). The plan reflected the concern among EPA research scientists that there was no safe level for dioxin in humans that could be determined from the toxicological studies. The upper limit for no dumping was set with regard to bioaccumulation time factors and capping effectiveness estimates that are described in more detail in the accompanying science note. EPA Region 2 adopted the proposal and submitted it to the steering committee.

The Corps immediately attacked the plan, however, saying that it was too conservative and also unrealistic. The New York State Department of Health methods were questioned. The method of applying these conservatively derived fish numbers directly to the bioaccumulation test species was questioned. The fact that the NYSDOH number was derived for a fresh water fish and applied to a marine environment was questioned. Phone calls were made to EPA management that painted a bleak picture for the viability of a number of important dredging projects were this plan to be implemented. Discussions continued, and there did not seem to be any light at the end of the tunnel.

Meanwhile, dredging projects that had done the testing required at the time of their applications were being held up for the Corps' decisions on their permits pending the resolution of the dioxin issue. The inability of the Corps and EPA to agree on a reasonable approach to address the dioxin issue was beginning to cause economic hardship to the shipping industries around the

The EPA proposal was arrived at by applying the 10 pptr NYSDOH guideline in a simplified risk-based approach. Critical information on long-term uptake (bioaccumulation) rates from the Narragansett study was used to make correlations from lab results to the ocean environment. In the study, bioaccumulation test organisms were exposed to a sediment that was fairly highly contaminated with dioxin (600 pptr of 2,3,7,8-TCDD—the term for the most potent congener in the dioxin family) over a long time period. The study was to determine the steady-state potential of dioxins, which is the length of time it would take organisms to accumulate as much contaminant as possible, or die trying. This is important to know because the standard laboratory tests for dredging projects should be designed to run for a reasonably short period of time. On the other hand, the same organisms living in the ocean can be exposed to the sediments for much longer. By knowing the bioaccumulation rate and how long it takes certain organisms to reach a "steady state" (where they will accumulate no more) concentration of a contaminant under a given set of conditions, an adjustment can be made to the dredging project test results to equate them to what would happen in the ocean. The Narragansett lab found that a commonly used test organism, a sea worm, accumulated dioxin from those sediment concentrations for up to about five months to reach steady state.

Figure 3 shows the first-order results of the study conducted by Pruell et al. (1993). The figure is of directly measured uptake in sea worms with time—PR is the symbol for Passaic River sediments, and CTL are the control results (pg/g wet wt means concentration by weight in the undried tissue of the organism in picogram/gram or ppt). TCDD is the most potent dioxin congener, which was the main one analyzed at the time, and TCDF is a similar but less toxic furan compound. (This figure is not in the Pruell et al. paper, which goes further into more technical issues relating to calculated bioaccumulation factors, but was allowed to be included here to more clearly illustrate the science.)

The sea worms accumulated approximately 2.5 times the amount of dioxin at steady state than after the twenty-eight-day standard test period. For the proposed plan, only the lower end of the uptake curve found in this study was applied, in concert with the capping management requirements. The plan had a requirement to cap a material within ten days, when only about a quarter of the twenty-eight-day concentrations had accumulated in the Narragansett study worms. The proposition was that a material that resulted in worm accumulations of less than 10 pptr after twenty-eight days would have little chance of resulting in accumulations of more than the NYSDOH level of 10 pptr in fish. This is because, if only about 25 percent of twenty-eight-day test accumulations would occur in ocean worms after ten days (before capping occurs), then material that resulted in less that 10 pptr accumulation in the test worms could result in only approximately 25 percent of that in worms or similar organisms exposed to it in the ocean site before capping.

This left a fairly good (conservatively, 3×) safety factor to compensate for the assumptions that (1) capping is probably less than 100 percent effective in the open ocean, and (2) there could be biomagnification of contaminants by fish that eat the worms. This safety factor was well within what is typically used in performing a risk assessment from toxicological data, where safety factors of 10× or greater are conventional for attributes such as differences in toxicity between species. There were no studies that reliably quantified the effectiveness of capping, since there were always too many variables or unknowns; however, it was generally believed that effectiveness was somewhere between 50 and 90 percent. Much depended on the accuracy of placement of both the dredged and the cap materials, the physical properties of the two materials, the oceanographic bottom conditions, and other such factors. The term "biomagnification" is used when a contaminant is accumulated to substantially higher concentrations in species higher up the food chain (fish) than in their food sources (worms). Although the scientific literature did not appear to indicate biomagnification of dioxin by the fish that were studied, its physical/chemical properties certainly indicated the potential. Dioxin has properties similar to those of contaminants such as PCBs, which had been shown to biomagnify. If a conservative estimate that the levels in worms could biomagnify to twice those concentrations in fish (a not unrealistic potential according to physical/chemical factors) is applied together with the probabilities regarding capping effectiveness, it can be seen that the 3× safety factor in the proposition was about right and not at all overconservative. (2.5 pptr × 2 × ~1.4 = ~7 pptr).

Port. There was a specific area delineated within the Port that was the "dioxin area of concern," based on sediment and marine life dioxin data. For materials dredged from within this area, dioxin testing was required. This limited area was specified, rather than requiring dioxin testing for all dredging and disposal projects in the Port, because of the high cost of laboratory dioxin analysis. EPA felt it had to proceed with some surety before requiring these greatly higher costs for all projects, so it required them only where there were data indicating the likelihood of dioxin contamination.

This "area of concern" also continued to be a point of contention between the agencies. The Corps wanted to limit the testing requirement to Newark Bay and its tributaries, the Passaic and Hackensack rivers. EPA felt there was sufficient justification from water current patterns and sediment movement

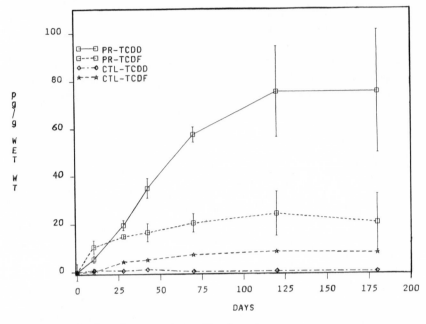

FIG 3. First-order dioxin bioaccumulation. (Pruell 1993)

studies for including the Arthur Kill and the Kill van Kull. (The reader will recall that these waterways are Newark Bay's links to the upper and lower New York Harbor.) The kills have important shipping channels and terminals that require periodic dredging. Most of the petroleum terminals reside in the Arthur Kill and the Kill van Kull. (The reader will also recall that the Port is the largest petroleum port in the country.) So, the inclusion of these two waterways in the dioxin testing area was no small matter for the petroleum industry, which needed to have its terminal berths dredged, and for the Corps, which dredged the navigation channels. In November 1990, NJDEP confirmed dioxin contamination in the Kill van Kull. The area of dioxin concern was thus expanded, and more projects were now subject to the dioxin testing requirement. This area of the Port is highly industrialized and shipping-intensive, and most of the major dredging projects are within it.

3

THE OCEAN CONNECTION

For some time before the dioxin deadlock, Clean Ocean Action and New Jersey congressman Frank Pallone were lobbying for new national ocean disposal legislation. Mr. Pallone, whose district was on the New Jersey shore in Monmouth County, worked with his delegation and was able to get a provision adopted into the Water Resources Development Act of 1986. The provision would keep EPA and the Corps busy doing ocean surveys for the next several years. The Water Resources Development Act is a major piece of federal legislation that is reauthorized every couple of years (usually even years) to fund all the large water projects in the country, including dam construction, flood control projects, and federal navigational dredging. Section 112 of WRDA 1986 required EPA to, among other things, designate an alternate ocean disposal site for dredged material at least twenty miles from shore.

This alternate site was to replace the existing Mud Dump Site, which as previously mentioned was about six miles offshore of northern New Jersey. EPA had officially designated the boundaries of the 2.2-square-mile Mud Dump Site in May 1984, but disposal of dredged material from the Port had been going on in the general area since the turn of the century. Disposal of dredged material from the Port had begun to take place two centuries ago and had progressively moved farther out toward the ocean and then into the ocean. There were obviously very few controls, or cares, about disposal of dredged material in the early part of the last century. It took the public until the middle part of the last century even to become concerned about disposal of sewage into the harbor, so concern about dredged muds from the harbor itself was minimal during all that time. Only with the advent of the Clean Water Act and the Marine Protection, Research, and Sanctuaries Act in the 1970s did any real focus on disposal into waters of sewage sludge, industrial wastes, and dredged materials begin to occur.

The public's alarm about the levels of dioxin and other contaminants that were ostensibly being dumped at the Mud Dump had also begun to grow in the late 1980s, mainly spurred on by Clean Ocean Action and some national environmental groups such as the Environmental Defense Fund, Greenpeace, and Coastal Alliance. It was Clean Ocean Action that really got the ball rolling, however, with the national groups providing some support. Clean Ocean Action was particularly effective in getting large signature lists of beachgoers, surfers, and fishermen. It was also becoming quite adept in other aspects of the art of public advocacy, engendering the support of local and state politicians.

Congressman Pallone was sensitive to the apparent widespread public interest, evidenced by the long signature lists that were presented to him by Clean Ocean Action. Many New Jersey shore towns are not only heavily dependent on ocean-based recreation for economic reasons, but since many owe their founding and continuing sense of community to being located on the shore, their connection to the ocean is almost spiritual. Residents of these shore communities rely in large part on the summer tourist business for their economic survival, and after the tourists leave in the fall and before they arrive in spring is their time to have the ocean to themselves. And the beachgoers that come to these shore communities to rent summer cottages or rooms at hotels during the summer months are often long-term annual visitors. It was probably not very hard to get the signature of almost every soul on a crowded summer beach to help fight the terrible Muck Monster (the descriptor used in COA flyers and on its website) that the federal government was creating just offshore from their beautiful beaches.

The fact that there was no evidence that material from the Mud Dump had moved anywhere near the shore was beside the point. As long as enough signatures were obtained from a concerned population, politicians could be moved to act. Clean Ocean Action probably knew that the real, or at least potentially real, risk of the dumping was to the marine environment immediately surrounding the Mud Dump Site, not to the Jersey shore bathers. Ocean current patterns had long been known to flow in a southerly direction offshore, and there was little likelihood that material disposed six miles out would migrate in any detectable amounts onto the beaches. But the population of fishermen and wreck divers that could be affected by pollution around the Mud Dump was much smaller than the population of beachgoers, and that limited population would have had much less public impact. Clean Ocean Action of course also got the hearts, minds, and signatures of commercial and recreational fishermen and divers as well as the beach community.

The political will was certainly there, if short of being able actually to stop ocean dumping of dredged material, to at least move the material a lot farther away from the Jersey shore. So Section 112 became law, funds were obligated, and a year or so later EPA began the process of scientific investigations for designating a suitable disposal site at least twenty miles from shore. Before it was all over, the costs for this work added up to about $1 million. Unfortunately, although this project made for some interesting oceanographic survey work for regional scientists, and some valuable oceanographic data were obtained, a disposal site for dredged material more than twenty miles from shore was never designated.

There was considerable resentment among the dredging community about Section 112, for two reasons. First, it was thought that the reasons for its enactment were purely political and had no scientific or environmental basis. More important were the added financial costs and risks to men and equipment of transporting dredged material that far out in the ocean. It was true that industrial waste, and later sewage sludge, was transported much farther out (106 miles) for legal disposal. But industrial waste disposal at such a distance from shore did not last very long before the message was perceived by the dumpers that going farther out in the ocean was not the way to go. There were cheaper ways, some subsidized by EPA, for making industrial production processes more efficient and minimizing or reusing waste products. As these were developed, ocean disposal of industrial wastes at the high costs being imposed was soon seen as a dinosaur. The sewage sludge issue was more influenced by state and municipal politics, and was more complicated in that sense. When EPA consent orders backed by the passage of the Ocean Dumping Ban Act started to inflict painful fines on municipal authorities, reassessment of disposal options took on new life. And when EPA-mediated efforts developed real and cost-effective alternatives for disposal and reuse of sewage sludge, the retreat from the ocean was a foregone conclusion.

But these two examples of deep ocean disposal could not be compared, in the dredgers' minds, to the situation with dredged material. Dredged material was not some toxic industrial waste product or even a human biological waste product. It came from waterways that were connected to the same ocean waters where the material was destined to be disposed of. It made no sense to the dredgers to have to expend unseemly amounts of additional fuel and time (dollars), not to mention the added risk, to dispose of this material farther out in the ocean just to quell some environmentalists' flighty trepidations. The Corps, PANYNJ, and others in the dredging community immediately began to lobby for the repeal of Section 112. Given the combined political clout represented in this lobby, it was not overly optimistic of them to think that they might have a chance for success. Taking what was thought to be the prudent course, the Corps decided also to fund oceanographic investigations of alternate sites closer to shore, in case the lobbying prevailed. As it happened, the Corps' presumption this time paid off.

There is a well-defined process included in the Ocean Dumping Regulations that EPA must follow in designating any ocean disposal site. First, a Zone of Siting Feasibility (ZSF) study must be done, in which all available information on a targeted ocean region is collected so that EPA can identify areas that have a better chance for eventual designation. This is considered to be a "tabletop" exercise using existing information and is mainly a process of elimination. Major shipping lanes leading into and out of harbors tend to be avoided. Military use areas generally tend to be avoided (the military is not overly sensitive about them, since they are just large areas demarcated for possible emergency restrictions). Any known popular fishing areas are definitely avoided. The regulations say that fish reproduction, spawning, or resting areas are to be avoided. However, there is usually either nothing known about these or they are known to be so vast an area that, unless better defined, they are usually ignored. For example, it is known that bluefish spawn over an area of the New York Bight that covers probably four-hundred square miles. If a two-square-mile disposal site were designated somewhere in that vast area, it would obviously not have a major impact on it.

This is true, of course, only on the condition that significant amounts of pollution do not spread beyond that two-square-mile area. (After the Hudson River Westway project debacle, environmental scientists in the New York area have used the term "significant" very sparingly because of its statistical/legal connotations, but sometimes trying to avoid using it becomes *appreciably* difficult.) The issue of contamination from a disposal site is extensively addressed in the Ocean Dumping Regulations. They contain provisions for taking concerted action to prevent pollution from dispersing out of the site if such a situation is discovered. These actions may include closing the site, regardless of the existing needs of the port that has to use it, and taking expensive remedial actions. Lastly, for designating a site, any other known oceanographic information, such as the physical type of bottom, the living marine communities on the bottom, and the current velocity and patterns, is looked at to try to narrow down areas that will then be further investigated by actually conducting ocean surveys.

It was decided that it would be necessary for the Section 112 site to be a containment site as opposed to a dispersive site. The intent here is as clear as it sounds; a containment site would tend to contain material placed in it; a dispersive site would disperse it (ostensibly in a more controlled and benign manner than portrayed by the old axiom "dilution is no solution to pollution"). The philosophy behind these two different types of ocean disposal sites can make sense, but as an astute marine scientist at the EPA Narragansett Lab, Walter Berry, would often say, the devil is in the details. For ocean disposal of say, liquid acid waste, dispersion was clearly the better choice. The acid would get neutralized by the ocean's natural buffering capacity, and any contaminants in the remaining liquid would theoretically be quickly diluted to imperceptible concentrations in the expanse of seawater. (The worst thing about the acid waste dumping was not the acids but the heavy metals that the liquid was contaminated with, which do not go away so easily and can contaminate the marine food chain.) The same strategy was applied for sewage sludge at the 106-mile site, where barges were required to travel predetermined lanes within the dumping area at a set velocity and discharge rate so that the liquid material would disperse as widely as possible. For dredged material disposal in this region, it was determined that the site should be contained.

But when you are talking about the ocean, containment is a relative concept. Unless the material that is dumped is actually contained in something, the coastal ocean is not an easy place to put anything into and then expect to find it again. Nevertheless, that is what is expected with this type of disposal site. Various factors such as the type of bottom and the bottom currents in an area can give an indication of how stable the bottom might be and therefore how well it can contain material placed there. Water depth is another important factor. Here is where WRDA Section 112 inadvertently blundered in the right direction. At twenty miles from shore and farther, the water is a lot deeper than where the Mud Dump Site is located, actually more than twice as deep. Depth is important because the greatest forces that can potentially disrupt bottom sediments are from surface waves, not from the typical currents in most oceans. Ocean currents are usually not strong enough to disrupt the bottom by themselves. Physical oceanographers know the exact relationship between water depth and the ability of surface waves to disrupt the bottom that most people would know intuitively in a general sense; the deeper the water, the less likely that waves can affect the bottom. What most people would not know intuitively is that it is not the height of the waves that makes the difference; it is their periodicity, or the distance between them. The relationship is that the bottom will "feel" the circular-motion energy (termed orbital velocity) from surface waves down to a depth of approximately one-half the wavelength (Leet and Judson 1965), the distance from crest to crest (or trough to trough).

Large storm waves in the open ocean, like those from a strong nor'easter, can easily exceed a wavelength of one hundred feet. That means that at a depth of fifty feet, some of the energy from those waves can begin to be imparted to the bottom. That does not necessarily mean the waves will be able to stir up and move the bottom sediments at that depth. Other factors regarding the type of bottom and its physical characteristics are very critical in this regard. For example, the grain size of the sediments, the presence of certain biological activity (which can either increase or decrease the potential for resuspension), and the degree of consolidation (or packing) of the sediments can all play a major role. As a good rule of thumb, though, other things being about equal, the deeper the water, the more likely the bottom will be stable and be able to contain what is placed there. So the fact that Section 112 required EPA to look at least twenty miles from shore for an alternate dump site resulted in some interesting findings.

The ZSF was done and four areas were targeted, two closer than twenty miles from shore (in deference to the Corps) and two outside twenty miles as required by Section 112. Next, ocean surveys to collect the additional necessary information on these areas were planned. These included side-scan sonar surveys of the bottom and setting out bottom-mounted cameras and current meters to record conditions on the bottom at the four areas. Current meters were also deployed at different depths, and wave-measuring buoys were set. Since the largest waves over longer periods occur during winter in this part of the world, a survey trip for deploying and maintaining these systems was not what most people would envision for a pleasant sea cruise.

The scientists and technicians who were involved in this work typically rotated on a four hours on–four hours off schedule, and could expect to work often at night on the exposed rear deck, many times in dead cold with rough sea conditions. Large "guard" buoys about five feet high and wide and weighing over half a ton were typically used to mark off and protect from passing vessels the array of oceanographic instruments tethered to the surface and reaching down to the bottom. The guard buoys were built to be able to withstand some abuse, such as a possible ship collision. The instruments on the bottom were mounted on quadrapods, four-legged pyramid-shaped heavy steel frames standing about six feet high and about six feet per side at the base. The quadrapods were tethered to buoys somewhat smaller than the guard buoys, since they only had to hold up the cable for the instrument array (which included current meters set at two or three different depths) leading to each quadrapod. An ocean survey was conducted every couple of months or so during the winter period to maintain and download data from the instruments. This entailed hauling the whole array of instruments and quadrapods onto the rear deck of the vessel in each of the four locations they were set.

The captain's modus operandi on these cruises was to back up to one of the buoys (for better navigational control), at night as it often seemed to be, and against large cresting waves driven by a stiff winter wind. Someone would then try to attach a winch cable to the instrument buoy with a long pole as it rose with a wave to within reach. Once it was wrestled onboard and secured, the winch cable was attached to the line for the instrument array and quadrapod. Then, as the huge main winch strained to bring the hulking quadrapod up onto the rolling deck, lines were secured to the quadrapod as quickly as possible as it swung around the deck to restrain it from crashing back into the boom or something less solid (like a scientist) and possibly damaging the sensitive instruments. During all this, the occasional outsize wave would suddenly appear out of the dark and break over the open transom, sending a torrent of cold seawater and spray over the work deck. The ship that was normally used for this work, the EPA Ocean Survey Vessel *Peter W. Anderson,* is a converted 165-foot Vietnam river patrol boat that rolls like a ball even on calmer seas, so off-duty time on one of these voyages was usually no picnic either (see figures 4–8).

The ocean surveys that were usually done during the summer, trawling up bottom fish and other specimens or sieving for benthic organisms in sediments brought up in grabs, were much preferred. But the physical oceanographic data from the winter cruises had to be collected first. A report on the physical oceanographic conditions at the Section 112 candidate ocean sites, including the ones more than twenty miles from shore, was completed in December 1990. The U.S. Geological Survey was contracted to help assess all this data because of its expertise and recent work in this area of the ocean. It submitted a report to EPA with a lot of technical scientific information and conclusions. The bottom-line conclusion was rather simple—the deeper the water, the more stable the bottom.

Nevertheless, much to the relief of the Corps and the other dredgers, Section 112 of the 1986 Water Resources Development Act was repealed in November with the passage of WRDA 1990. EPA could now look for sites closer to shore. In place of the old Section 112 was promulgated Section 412 of WRDA 1990, which included requirements that reflected the new direction that the environmental groups, and the government agencies to some degree, felt was the better way to go for managing material dredged from the Port. Section 412 required that within 180 days EPA and the Corps would report to Congress on (1) the sources, quantities, and characteristics of the Port sediments, (2) measures to find alternatives for ocean disposal, (3) measures to reduce quantities of dredged material for ocean disposal, (4) measures to reduce contaminants in dredged materials, and (5) a monitoring program for the ocean disposal site. A major effort was also required to conduct a nonocean alternative disposal demonstration project that could include borrow pits, a containment island, landfill application, habitat restoration, or decontamination technologies. Along with the requirements, a generous funding allocation of $4 million was included to carry out these requirements, $3 million for the first parts and $1 million for the demonstration project.

It was decided that decontamination technologies would be pursued to fulfill the demonstration project requirement. No small factor in this decision was that Clean Ocean Action and Congressman Pallone were in favor of decontamination and were influential in the congressional funding allocation for the act. This did not mean that the agencies had not been looking at decontamination technologies prior to 1990. EPA and Corps offices in the Great Lakes Region had been investigating decontamination technologies and their associated issues in a program called ARCS, Assessment and Remediation of Contaminated Sediments. This had been started a few years earlier to address the many contaminated sediment problem areas within the Great Lakes. The findings that resulted from that effort were applied to planning a New York program to address WRDA 1990.

Although there was a search for an alternate disposal site going on, EPA also had the responsibility of keeping track of the conditions at the Mud Dump Site. The Corps of Engineers had been taking on the responsibility and financial burden of monitoring the site for the past years. The main things that the Corps was looking at were changes in the bottom topography, which it did by conducting bathymetry, or depth, surveys. These kinds of investigations provided information on the things that the Corps was most interested in. When the Mud Dump Site was designated, a requirement was that no more than 100 million cubic yards of material could be disposed of there. This was mainly for navigational safety, calculated to leave the bottom deep enough for safe navigation by ships that might stray from the main sea-lanes. The Corps had to keep track of how fast the site was increasing in height. Another reason to monitor the bottom topography was to check on the amount of material that the dredging companies had actually dumped at the site.

Given adequate sensitivity of bathymetric methods, the volume of material added to the site could be calculated and also compared with the amount that had been contracted to be dredged and disposed of. This could answer the question, were the dredgers hitting the disposal site or short-dumping along the way? Short-dumping was a known transgression that had sometimes been committed by dredgers because of the ease and economic advantage of doing so. It involved dumping a load at a location or along a route closer to shore instead of actually steaming six miles out to the disposal site. Historically, this could mean shorter than whatever distance was required in times past. The general area had been used for dredged material disposal since the turn of the century. During the early and middle 1900s, dumping first occurred usually within the harbor itself and was gradually moved farther out into the ocean as there developed increasingly less tolerance for it. In those times, though, the dumping still had virtually no restrictions on how or specifically where it occurred. By the time the six-mile site was formally designated in 1984, dredgers were required to dispose of materials within the confines of the site, according to buoy locations and navigational location. But there occurred over the years from time to time instances of short-dumping along the way out to the ocean site, and dredgers had been cited for it by the Coast Guard and the Corps. The usual reasons for short-dumping were savings on fuel costs, taking shortcuts during tight schedules, and sometimes foul weather (the only potentially allowable excuse, under conditions of threat to life or vessel). All the reasons listed above for doing bathymetric surveys applied both to regulated permit holders as well as to dredgers doing the work for the Corps' federal projects (federal channel authorizations).

Regional EPA staff (author) believed that it was past time to conduct more intensive monitoring than had been done before at and around the Mud Dump

Site. Chemical analyses of the sediments and of bottom organisms in and around the site needed to be done to better assess whether the site was causing potential environmental harm to the nearby marine ecosystem. EPA management became convinced this was necessary (despite Corps opposition), and EPA funding was obligated for a chemical/biological survey and full data analysis. The most expensive part of this kind of study is usually the chemical analysis of the large numbers of samples that are taken, both of sediments and of benthic (bottom-living) organism tissue. The cost for this study was more than three times higher than for the typical bathymetric surveys the Corps had been doing, but the EPA study proved to be well worth its cost. The full focus of the environmental groups had not yet turned to questions about conditions at the ocean dump site. They would soon be asked, though, and the results of this first biological/chemical survey were available by that time to address the very pertinent and potentially embarrasing questions that were raised. EPA, at least (the Corps may not have cared much at the time), would have been left looking rather poorly if it had not yet collected the kinds of data and information that were being considered as basic needs.

The survey was completed, and the results were finally made available. The analytical testing that was done for the survey was actually more extensive than that done for testing dredged material then and in the recent past, so it was not really known what the findings might show. The testing protocol at that time had required chemical analysis for only a few metals in sediments and only a few additional compounds in bioaccumulation tissue. The chemical analyses done for the survey samples of sediment and tissue included many more contaminants that had not been analyzed for during dredged material testing. So, given the recent findings on dioxin and the growing concern regarding sediment contamination in the Port, the worst was feared. After years of disposal of dredged material from one of the most polluted ports in the country, anxious regulators were hoping for the best but not really expecting it. There was therefore considerable relief when the biological and chemical monitoring results indicated the site was, if not exactly clean and pure, not a horribly polluted area ridden with toxic muds and mutated life-forms. The survey showed that there were locations where sediments and benthic organisms had fairly high levels of contamination and other areas that were relatively clean, even within the dump site.

The more contaminated areas were typically where the most recent dumping had occurred within the Mud Dump Site. Even before this time in the late

1980s, there had been sections that were placed off-limits for dumping owing to the desire to not disturb an experimental capping project done in the early 1980s. This area and some others that had not been dumped on for the last five or more years showed fairly uncontaminated conditions, and the earlier capped area still had a significant amount of sand on the surface. (The sand may have remained from the sandy cap material used, or could in part have been sand that had drifted in from the surrounding area—the survey was not designed to address this issue.) The more recent disposal areas generally had the highest contamination levels, although even these were not as bad as some had feared—concentrations of contaminants were not as high as in some of the most badly polluted sediments in the Port, so the news could have been worse. Later, other surveys would be done that used different sampling and testing methods, and these would result in a darker picture of the environmental conditions in and around the Mud Dump Site.

4

REVISING REGULATIONS

TESTING, TESTING . . .

It was February 1991, and the Corps and EPA had decided that a good place to hold the annual ocean dumping coordinators' meeting would be Monterey, California. Whatever else could be said about their headquarters (and a lot was), the regional ocean dumping coordinators were unanimous in their regard for headquarters' ability to pick outstanding places to hold the national meetings. No dreary business district or forgotten backwater for them—a nice place by the ocean was usually found. Ocean dumping coordinators' meetings were held in places like Annapolis, Monterey, Pensacola (Pensacola may not be glamorous, but it *is* on the Gulf Coast). After all, it would not do to discuss issues of national coastal importance in places that were not suitable to promote the right frame of mind. The Monterey location was in a beautiful, albeit older, hotel-conference center built on a rocky shore of Monterey Bay. The nearby sea lions, sea otters, and kelp forests (along with an occasional rental kayak ride for a closer view of same) helped to promote the right frame of mind.

The main subject of discussion was the implementation of the revised Green Book. The issuance of the national testing manual had become a subject of growing interest among the agencies and the environmental community, but previously projected release times for it had been passed because of unresolved technical differences between the Corps and EPA. The EPA Headquarters office now promised it would be officially issued in two weeks. The regions would have to prepare their regional manuals in accordance with the revisions and implement them within six months. The EPA regions and most Corps districts did not seem to have a problem with this schedule. The New York District Corps representative did. His concern was that it would take longer than six months to adequately revise quality assurance measures under existing contracts for laboratories doing the Corps' federal dredging project testing. Then one or two other districts agreed with this point or raised a similar issue. There was considerable discussion in breakaway sessions on this issue, but no resolution was achieved at the meeting, which was a sign of the continuing difficult negotiations that lay ahead. Implementation of the EPA Region 2/New York District regional manual would be a continuing struggle for some time to come.

Dioxin continued to be a major part of the discussions regarding testing of dredged material for ocean disposal. At first, the question was whether even to include a requirement for dioxin testing as part of the battery of tests that were normally required for applicants. This issue continued to revolve around whether a dredging project was, as mentioned before, within the "dioxin area of concern." By agreement, and because of the high testing costs for dioxin, EPA and the Corps would require the testing only if there was reason to believe that dioxin might be present at levels of concern in the project area. As more information came in from various sources implying the presence of dioxin virtually throughout the Port, the area of concern would expand eventually to include almost all dredging projects. For this time period, however, the scope of dioxin testing was a subject of continuing controversy between the agencies.

The revised Green Book (U.S. EPA/Corps 1991) was issued in June 1991. Although methods for testing and guidelines for how to evaluate test results are quite explicit in most cases, there were remarkably many details that needed interpretation when it came time to implement a regional testing manual. For example, a major change in the toxicity tests was the requirement to use amphipods. An amphipod is a small crustacean that looks a little like a shrimp, but most species are only slightly larger than a pinhead, a few millimeters long. Sediment scientists had been studying them, along with other benthic organisms, for their potential as sensitive indicators of contaminated sediments. By sheer weight of scientific evidence and predominant opinion in the scientific community, Corps scientists at WES agreed with EPA that this was an appropriate organism to adopt into the new guidelines. Considerable testing by EPA and other scientists had indicated that amphipods were sensitive to contamination but were not the most sensitive of all possible organisms. As test animals, they did after all have to be able to withstand some degree of handling. Also, they would have to be able to handle other testing-related conditions not related to contamination, such as varying amounts of turbidity or grain size or pH associated with different dredge material samples.

An issue about the amphipod tests that was left somewhat open in the Green Book was brought up a number of times and had to be dealt with—whether a flow-through or a static test should be done. This was an important consideration since the results could depend greatly on it. In static tests there is no water change during the test period, whereas in flow-through tests water can be exchanged into and out of the test chambers at prescribed rates. When some dredging projects that had not failed the earlier toxicity tests began to fail the amphipod tests (which were being conducted under static conditions), questions began to be asked by the Corps and others about the test conditions. Perhaps a static test, in which the water is not changed over ten days, results in toxicity that is unrelated to contamination. Perhaps a buildup of ammonia or other toxicants resulting from decaying organic matter in the sediments was causing the organisms to die, not the chemical contaminants that were present. Ammonia was claimed not to be a relevant toxicant for ocean disposal testing because in actual dumping operations it would harmlessly dissipate from material dumped into the ocean.

The 1991 Green Book states (although the guidance is somewhat conflicting in different sections) a preference for flow-through tests, if ammonia or other factors dictate, but prescribes static systems in most cases. The draft New York regional manual required a static test since that was the procedure that had just been standardized by the American Society of Testing and Material (ASTM). The static tests require putting an inch-or-so-thick layer of the sample sediment in enough separate aquaria (usually glass beakers) to provide the required number of replicates for each test and reference sample. Then several inches of seawater are gently poured in to cover the sediments, and finally twenty test animals are put into each aquarium. The procedure calls for the system to remain undisturbed for ten days, after which the numbers of live organisms are counted. There are all kinds of explicit requirements for each of these seemingly simple steps to ensure consistency of the testing procedures. The pass/fail point in the tests is mortality that is statistically significant and more than 20 percent higher than for a reference sediment. Reference sediments are supposed to be relatively clean sediments collected from an area that is near the intended disposal site but has not been affected by any prior dumping.

There was a hue and cry from the New York dredging community to change the tests to allow for water renewal. The problem that the regional EPA and

When sediments that contain ammonia are dumped in the open ocean (or any other large body of water), the ammonia might readily dissipate into the surrounding water during the dumping process. In a large, nonstagnant body of water such a release of ammonia would not be likely to have much adverse effect on the environment, because the ammonia would quickly be diluted to harmless concentrations. Whether the process would occur adequately, or harmlessly, would depend on the specific conditions of the waterbody and the amounts and characteristics of the dumped material.

The quick dissipation described above would be more likely with unconsolidated, looser material, which would quickly mix with seawater during disposal and lose much of its ammonia load. However, there could be a different scenario. A project might be dredged, for example, using a large clamshell bucket (the most common equipment used in the Port; see figure 9) of 10 or more cubic yards. This bucket would grab a large volume of shallow and deep sediments in one "bite." Deeper sediments are more likely to have higher concentrations of ammonia, from anaerobic decomposition that takes place below the oxygenated layer. The huge bites taken by large clamshell buckets can remain quite intact after placement in a barge and transport to the disposal site, depending on how compacted the sediments are and other physical factors. A split hull barge would release these large clumps where they could still remain largely intact even after contacting the bottom.

Under such conditions there would be little release of ammonia to the water column during disposal relative to the volume of ammonia remaining in the material on the bottom. Since large clumps of dumped sediment can be several feet or more high, they provide structural relief to a sea bottom that is normally flat and sandy in disposal areas. These clumps can act as attractants to fish and the benthic organisms that they feed on. The high-ammonia-content sediments that were buried three or more feet deep in the Port, out of range of most burrowing organisms, would now be readily exposed to colonizing organisms and fish at the dump site. So the decision of whether ammonia is or is not a contaminant of concern for ocean disposal was not an easy one.

other parties had with allowing water renewal was that other contaminants that were of concern could also be flushed away along with the ammonia.

The problem of whether or not flushing should be allowed during tests was one that tormented quite a few scientists in several agencies for some time. At first, EPA researchers who had been prominent in developing the tests were adamant that the static tests were fine. All the research seemed to show

SCIENCE NOTE: UNDISPERSED CONTAMINANTS

The amphipod tests, after all, were supposed to evaluate the potential for adverse effects from contaminated sediments on bottom-dwelling and other marine organisms. These potential effects have been assumed by researchers to be caused not only by contaminants bound to the sediment particles but also by contaminants more loosely held in the "interstitial water" of sediments (sediments contain varying amounts of water depending on many different factors, the most important being depth—typically, the closer to the sediment/water interface, the more interstitial water present). These contaminants, including heavy metals, pesticides, PCBs, and dioxin, are not dissipated as easily as ammonia, and if they were, that would be a problem. Ammonia is quickly oxidized to harmless components after contact with oxygenated waters such as seawater—but other contaminants released during disposal, such as those listed above, could create a hazard to the receiving waters. It's more likely that when contaminated sediments are dumped into the ocean, much of their contaminant load remains after the sediments contact the bottom. These undispersed contaminants can have toxic or cumulative effects on colonizing organisms similar to the test animals. So if the test conditions were such that these contaminants were dissipated to overlying water much more thoroughly during the tests compared with what would happen in the ocean, the tests would not be sensitive indicators of the potential adverse effects, irrespective of how sensitive the test organisms were.

that only sediments that were truly contaminated caused test failures. But the opponents were persistent. Corps Waterways Experiment Station scientists presented data that showed that ammonia at levels sometimes measured during dredge tests could be toxic to amphipods. On the other hand, information came in from other EPA regions around the country using the tests that did not find ammonia to be a problem. To flush or not to flush continued to be a major issue for a good while.

There were other technical issues that would provide for continuing negotiation, recrimination, and all-around good fun between EPA and the Corps. One of these was how to assess the potential for adverse effects from the dispersion of sediments and their contaminants to the surrounding ocean waters during disposal. The concern here was effects on ocean waters and their marine resources, not on the benthic environment. It is widely believed that the greatest potential for adverse effects from ocean disposal of dredged material is to the benthic environment, since dumped material settles to the bottom quickly and generally stays there. However, there can also be impacts on the water column, and the Ocean Dumping Regulations require evaluating the potential of adverse impacts on this part of the ocean environment. Though most of the material dumped during a disposal event will get to the bottom quickly in a mass, depending on its physical characteristics a significant amount of the finer and looser (not cohesive) sediments will remain suspended in the water column for longer periods. It is these finer organic sediments that can actually hold most of the contamination in a load of dredged material, because of physical/chemical binding factors, so this potential exposure pathway also needs to be evaluated.

The preexisting procedure used a simple mathematical mixing model that applied information on water depth, currents, and some limited material testing information (physical and chemical) to estimate the potential effects to the water column. The 1991 Green Book contained and recommended a much more complex model that required more information on the ocean site and the material to be disposed of. The model used a complex system of hydrodynamic mixing equations and input parameters that required very specific oceanographic information on the disposal site and detailed chemical/physical information on the proposed dredged material. This level of information had not been collected prior to 1991, nor had a model of this complexity ever been required to be used by regulators doing the evaluations. The outputs of the model are of two general types: one calculates the concentration of key contaminants projected to occur at the boundaries of the disposal site, and the second estimates the toxicity of the disposed material, also at the disposal site boundaries.

The first type of calculation uses as input the results of chemical analyses of the sediments, and the second uses the results of suspended-phase toxicity bioassays; both of the above are applied to the mixing equations and disposal site parameters of the model. The results answer the questions of whether the proposed material disposal would violate marine water quality criteria (for the first case) and whether the disposal would violate water column toxicity criteria (for the second case). The regulations require that both of these assessments be applied at the boundaries of the disposal site. Workshops had been given around the country by a Waterways Experiment Station specialist on the model, since it was anything but user-friendly for the regulatory scientists that had to use it. Also, others who had tried using the model discovered that there were still bugs in it. The implementation of this model in any region was therefore not a certainty, and the New York District Corps wanted to leave its options open as long as possible. (Jumping ahead a little, the bugs were eventually worked out, and the model was applied by New York District Corps regulators for the first project that broke the crisis logjam, the East River federal navigation project.)

Another issue that continued to be a source of disagreement between the agencies was related to the above mixing model but had repercussions far beyond it. This involved whether or not to conduct chemical analysis of sediments as part of the battery of required tests. The Corps had its reasons for not wanting to include this analysis, mainly revolving around the fear that such information would be misused by the uninformed public. There were beginning to be developed around this time several versions of sediment quality criteria, which are numerical standards that can be directly compared with a dredging project's sediment chemistry results to evaluate its potential for adverse effects. More will be said on this later, but suffice it for now to say that the Corps had grave misgivings about these standards for two main reasons. One was that it believed the sediment quality criteria being developed then were derived by methods that were too conservative and so the values were oversensitive, and second it believed strongly that the principle of bioassays as ultimate determinants of a sediment's toxicity would be undermined. The Corps felt that bioassays—testing for effects on organisms—were the best way to accurately gauge the potential for toxicity in individual project areas. The Corps was not alone on this since most EPA experts also thought bioassays were a better final determinant of potential adverse effects—the main difference being that EPA felt that appropriate sediment criteria could play a role in the evaluation process. EPA was not quite as given to the slippery-slope concern as was the Corps. One of the mixing model's requirements, in any case, was to evaluate the material for the most likely problematic contaminants, and this logically required a knowledge of the material's chemical composition. There are other reasons, discussed later, for conducting this type of analysis. The Corps' recalcitrance on this issue continued for quite some time (although, as in the above case, the sediment analysis was eventually fully implemented).

One of the most important testing issues that was a source of continuing conflict was the bioaccumulation test. As mentioned earlier in regard to dioxin, these tests measure the amount of a contaminant that test organisms take up in their bodies. The existing bioaccumulation tests were rudimentary in that they analyzed for only four contaminants, whereas the revised procedures called for increasing this number more than tenfold and for using more sensitive techniques. Concentrations of the contaminants found in test organism tissues are compared with concentrations that have been estimated as having a potential to cause adverse effects to the test organism or its food chain. The specific effects and the concentrations that are calculated to be associated with them are derived usually from other kinds of laboratory tests designed to measure the effect of a contaminant on test organisms, or from observed effects in the environment. The results of these kinds of "exposures" are applied in an approach that takes into consideration the actual exposures in the field, differences among organisms, and other factors to estimate an appropriate tissue concentration threshold, or "effect value," for the dredged material test organisms. These effects values, though, as will be discussed more later, were not widely agreed on and hence the subject of continuing controversy.

The long-held and continuing District Corps' position was that it would be inappropriate to implement the new bioaccumulation tests unless and until effects values with which to evaluate the results for all tested contaminants were mutually agreed on. EPA's problem with this stance was that not doing such tests after it became known that a wide range of contaminants existed in Port sediments would be ignoring the potential for the release of harmful concentrations of contaminants in proposed dredging projects. There were some generally agreed-on upper levels of concern that could be applied for some contaminants if test results were quite high, for example, and in any case bioaccumulation results could be relevant in making a final determination when toxicity test results were inconclusive. (This therefore appeared to EPA to be an irresponsible position taken by the Corps—not having agreed-on standards for all contaminants should not mean there was no reason to know what contaminants would be accumulated and in what concentrations.) The need to evaluate for bioaccumulation effects was also stipulated in the regulations. This issue would become the most controversial of all of the testing revisions, and it would play a major role in the continuing crisis.

5

SEEKING ALTERNATIVES

Of all the possible alternatives to ocean disposal that were ever considered, borrow pits in the Port were the top choice in almost every evaluation category. For their proximity, apparent practicality, low cost, and potentially short implementation time, borrow pits were the all-out winners. With regard to the other two main evaluation categories, environmental and public acceptability, pits were much more of a mixed bag. Subaqueous borrow pits, it will be remembered, are depressions that remain from past sand-mining operations in several locations throughout the Port. Calling them pits is somewhat of an exaggeration since, although they are deeper than the surrounding harbor bottom, some are much more wide than deep and their side slopes are in some cases fairly gentle. Alternatives for ocean disposal had been thought to be needed mostly for the more contaminated dredged materials, since there was not much concern about "clean" sediments going to the ocean. Borrow pits had been considered viable options for disposal of more contaminated sediments because material placed in them could be confined and capped with cleaner material, and thereby be segregated from the marine environment. There were also indications that some pits had lower oxygen and higher contaminant levels than surrounding bottoms (because of reduced water flow and increased sedimentation) and could perhaps be considered areas that were already contaminated. So the Corps believed that the public concern about disposal in borrow pits could be mollified if these concepts were presented in a public meeting. That was the Corps' first mistake.

In a public meeting that was held in the late 1980s by the New York District Corps of Engineers in Brooklyn, the magnitude of this misinterpretation became painfully obvious. The Corps presented its findings and conclusions, which included all the above points, illustrated with data, maps, and charts. Local fishermen claimed that the pits were not environmentally degraded but were instead great fishing holes, and they had unkind words for the representatives at the podium. A scientist from the National Marine Fisheries Service had supported the Corps' proposal based on findings that some of the pits had degraded conditions in them, and he said as much to the audience. The fishermen took offense at this, coming (they felt) from a member of the government that was supposed to be protecting their fisheries. He was escorted to his car after the meeting by New York City police officers, who feared that he faced a realistic threat of bodily harm. Local residents and their representatives voiced concern that the "toxic sludge" would escape and ruin their waters and shores. Far from being the great hope as an alternative to ocean disposal, borrow pits were turning out to be somewhat problematic.

It is curious that the Corps did not have a better appreciation before the meeting that there was strong public discomfort about the borrow pit proposal. It had previously met with elected officials in Staten Island and Brooklyn, and had heard from other public representatives and citizens before then. It knew that there were concerns among the public about the plan, but it underestimated how virulent they were. The Corps may have believed that its presentation at the public meeting would convince people of its technical expertise and operational abilities, but that was not the case.

When the Corps went back to the drawing board and tried to figure out where to go next, EPA and others in the dredging community commiserated with it. EPA felt the Corps' pain, but at that point in time EPA's thinking was, it's the Corps' responsibility to worry about disposal alternatives; EPA's responsibility lay only in deciding on what could or could not go into the ocean. That was EPA's first mistake.

In June 1991 regional EPA Water Division director Richard Caspe attended a meeting sponsored by the National Marine Fisheries Service at its laboratory compound on Sandy Hook, New Jersey. Scientists from NMFS had been monitoring the conditions at the old 12-mile Sewage Sludge Site. Recall that EPA had required sewage sludge dumping to halt at the 12-mile site by the end of 1987 because of the polluted conditions that were being found in the area and had moved the dumping to a site 106 miles offshore. NMFS had taken an interest in seeing how the 12-mile site would recuperate after the dumping stopped. After a few years of monitoring, researchers found that the area was actually recuperating very nicely, much faster in fact that anybody had expected. By all measures, the site was quickly returning to normal oceanic conditions, at least as normal as can be expected for a prior sewage sludge disposal site just offshore from a major metropolitan area. That was one message. The other message was not a good one. NMFS was finding evidence that contamination from the Mud Dump Site was being found in an important area that lay between it and the old 12-mile Sewage Sludge Site.

This area is called the Hudson Shelf Valley, a unique feature of the coastal ocean with importance for regional fisheries. It is the geologic extension seaward of the Hudson River, carved into the continental shelf during the last ice age when sea levels were much lower and the area was dry land. It meanders eastward about one hundred miles to the shelf break, where the drop-off to the Atlantic Ocean abyssal plain starts. (If you turn back to figure 2, the Hudson Shelf Valley can be seen as the dark set of depth contour lines running out of the Port toward the bottom right. The contour lines fade in the area midway to the shelf break, meaning that the depth differences are not as great, then pick up again as the break is approached. The feature is nonetheless a consistently deeper area all the way out on the continental shelf.)

Though the Hudson Shelf Valley is not a sharp geologic feature in all locations, having filled in to some degree with sand and sediments over time, it still presents a relatively significant drop in the surrounding flat topography of the coastal shelf. This makes it attractive for some forms of sea life and therefore is an area that exhibits fisheries resources including lobster, bottom fish, and pelagics such as tuna and sharks. Bill Figley, longtime manager of New Jersey's artificial reef program and fisheries expert, notes that some fisheries biologists believe that the Hudson Shelf Valley also acts as a separator of populations of fish species on the continental shelf. This can have important connotations with regard to fisheries resource assessment and planning. At its head, just outside the entrance to New York Harbor, is a large bowl-shape depression on the seafloor called the Christiainsen Basin.

The Mud Dump Site and the old 12-mile Sewage Sludge Site are located at opposite ends of the basin. One of the main reasons for closing down the 12-mile site was that sewage sludge was heavily contaminating the Christiainsen Basin and Hudson Shelf Valley. Sludge and its by-products would tend to settle into these more quiescent depths, and the resulting highly organic sediments were causing low dissolved oxygen conditions that could affect a large area during summer periods. These conditions had in fact been recorded in some years, sometimes resulting in massive kills of marine organisms. Another problem was that there were chemical contaminants associated with the sewage (as there always are), and these were accumulating in marine organisms living in the area. Lastly, the problem of microbial pathogens from the sewage was also a concern, with the potential to harm both humans and the marine food chain. At the same time that NMFS scientists were happily finding that the Christiainsen Basin and Hudson Shelf Valley were cleansing themselves of the effects of sewage sludge dumping, they also found evidence of contaminants from dredged material there.

This was not good news to the EPA division director. Besides the black eye that it gave the agency regarding its site management responsibilities, the Ocean Dumping Regulations are fairly clear on what has to be done in this kind of situation. The options are pretty drastic, including, if deemed necessary, closing down a dump site and starting remediation procedures. When Mr. Caspe returned to the office the next day, he initiated a set of actions to deal with the situation that would put EPA into a much more proactive partnership with the Corps in managing the Mud Dump Site. Meetings with the Corps' Operations Division followed, and a number of measures were discussed to tighten permit requirements and surveillance of operations, and to use updated navigational controls. These were designed to ensure that material was dumped in areas of the site well inside its borders, and especially away from its borders with the Christiainsen Basin.

The theory was that material was escaping from the site boundaries because of sloppy dumping practices. Calculations were made of the amount of material that might have slopped across the site's northeastern boundary into the Christiainsen Basin. These were based on data from new surveys that had been conducted to measure the height of mounds in the site including along its borders. It was found to be a huge amount of material. It was also resolved that a better management plan for the site would be developed, with EPA having much more control over it than in the past. The pressure for finding alternatives to ocean disposal was cranked up a few notches.

With the requirements of WRDA 1990 and the recognition of the long-range nature of the problems at hand, the New York District Corps Planning Division was becoming more involved in the dredging program. Meetings between EPA and the Corps on dredging issues had almost always been with the District Operations Division, as it was primarily responsible for the permitting and ocean disposal program. Now EPA met with Corps Planning people on specific WRDA issues, and the discussions were sometimes frank and broad-ranging. There was a general understanding that contaminant levels going to the Mud Dump had to be greatly reduced, or the ocean would soon be shut off for *any* disposal.

An idea proposed by the Planning Division director at one such meeting was for a large containment island in the ocean that would be a codisposal operation. Dredged material would be mixed with sewage sludge and perhaps other waste materials such as incinerator or coal fly ash, thereby providing a solution for more than one waste stream. He said he had had positive discussions with the Interstate Sanitation Commission, which had interests in most or all of these waste streams. Although EPA acknowledged that this was potentially not a bad idea, it also posed a lot of problems. One of these was that WRDA 1990 did not specifically provide funding for this option. It was noted that it was OMB's responsibilty to provide funding for projects of this kind. What was unsaid was that it was just too big an idea—the scope of the project was too huge to contemplate.

Each of those waste streams had its own set of laws, regulations, and programs established to deal with the different sources and characteristics of the materials involved. To try to coordinate such an endeavor among all the diverse programs and various interests would be an undertaking that was beyond the imaginings of the EPA managers. Though it was true that OMB would have to approve funding for such a project, identifying needs and programs are nevertheless within the purview of agencies such as EPA and the Corps. The growing political intensity and technical difficulties of the dredging issues, however, as well as the need to deal with the specific requirements of WRDA 1990, no doubt left EPA management with little appetite for venturing into grand new directions. Of course the Corps planning people could have gone to their headquarters on their own (and may have), but these other waste streams were EPA's responsibility, and without EPA support it was unlikely the plan would have gotten far.

6

SCIENCE AND POLITICS

The amount of time that was going by and the number of projects that were being held up for decisions were growing steadily. In an August 1991 ocean dumping coordinators' meeting held in Pensacola, an amphipod laboratory was set up so the regulators could see the lab tests themselves and understand the methods for getting the best results. Although the tests were standardized and most labs could get acceptable results, some were having problems. It was stressed at the workshop that developing expertise in manipulating the organisms during field collection and in the laboratory could make crucial differences in the test results. EPA bioassay scientists gave talks on the latest findings and demonstrated the best way of getting good, consistent results. Getting good results meant achieving acceptable survival rates in the controls and reasonable consistency between replicate test chambers. (Controls are clean sediments having the right physical characteristics for the test organisms, in which they should be quite happy and survive well.)

Both control sediments and reference sediments are required to be used in the bioassays. Control sediments are used to evaluate the overall health of the test organisms—if they do not do well in these, a failure of the test sediments will be suspect, either because of the organisms' poor state of health or because of some other aspect of the tests. The reference sediment has a somewhat different purpose, which is to produce results from relatively uncontaminated marine sediments that can be compared with results from the test sediment in order to make a regulatory determination. Reference sediments are to be collected from an area close to the ocean disposal site but not affected by disposal at the site. They should thereby be no more contaminated than the general surrounding area of the site, which, it will be recalled, should not have become contaminated if the site has been managed properly. For the amphipod test, there has to be a statistically significant and greater than 20 percent difference between the reference results and the test sediment results for a sediment to fail.

The test and control aquaria are typically set up as five replicates of one homogenized sample, which means they should all be very similar in composition. Large variations between replicates in results will therefore mean something is wrong. Twenty labs around the country were now getting consistent and well-controlled results with amphipods on a variety of sediments. One species was being cultured in labs on the East Coast so that there would be a stable alternative to collecting specimens from the wild or purchasing them from the only vendor at the time, located on the West Coast. EPA Region 2 was doing everything that it could, beyond what was apparently necessary in most other regions around the country, to help its own regional laboratories become proficient in carrying out the tests.

To some in EPA, it seemed that on certain issues the Corps had philosophical blocks that overshadowed a sound technical argument. One of these was the Corps' reluctance to do, or to require its regulatory applicants to perform, sediment chemistry as part of the tests. As mentioned before, its main objection was that the public would focus on the sediment chemical results more than on the bioassay tests, which the Corps feared could make projects look worse then they really were. The Corps placed little faith in sediment "criteria" or guideline values for contaminants that had been developed by several different groups. Sediment guideline values had been proposed by a Canadian government environmental agency, by a pair of NOAA scientists (Long and Morgan 1991), and by several state environmental agencies. The methods employed in these efforts used generally similar approaches, relating measured levels of sediment or tissue contamination to various observed effects in different species. All these groups presented their proposed values as guidelines, not necessarily as criteria for regulatory decisions, but it was realized that once numbers were issued they may tend to be referred to for that purpose.

EPA was also working on sediment quality criteria (and called them that) for five organic contaminants, and the Corps lost no opportunity to criticize the science behind their development. It felt that the methods were not scientifically sound, mainly because there was too much uncertainty in the assumptions that were typically used.

It was true that some of the various criteria or guideline values that had been made available (EPA's would not be issued for several more years) were quite conservative. Most were purposely set at quite protective levels so that there would be few false negatives, that is, tests that would incorrectly indicate a low potential for adverse effects. (A false positive, on the other hand, would occur if a guideline was set so low that almost any detectable result would indicate a potential for adverse effects, even if that potential was drastically remote.) It was generally acknowledged that there was uncertainty in the values, so all the sediment chemistry guidelines were proposed to be used as screens for potential effects, not as final arbiters for regulatory decisions. If the guidelines indicated a finding in the positive for potential effects, that might call for further evaluation by other means, such as bioassay tests. The Corps was concerned that people would want to use the sediment guidelines instead as strict pass/fail criteria and therefore was dead set against them.

The revised Green Book was fairly clear on this subject, however, linking the importance of sediment chemistry to identifying the proper contaminants to analyze in bioaccumulation tissue and to identifying the appropriate contaminants for application of the mixing zone model, and also noting its importance as a quality control check. There were many back-and-forth arguments on this issue, as there were with all the other testing issues mentioned earlier. In the end, all the contested procedures would be agreed to by the Corps when the regional implementation manual was finally issued on December 18, 1992.

Besides the testing issues that were being negotiated, management of the ocean dump site was also changing. Surveillance of ocean dumping operations was a topic that neither the Corps nor the dredgers were very enthusiastic about. They knew that any required changes in their procedures would almost assuredly be directly translated to increased costs. It was clear, though, that measures would need to be taken, and they were. Additional EPA staff were assigned to work with the Corps in managing disposal operations. Electronic surveillance equipment, known commonly as "black boxes," was required to be installed on barges to record the precise time, location, and height of the barge in the water. This way, regulators could tell exactly when and where a load had been dumped, because when a split hull barge starts to dump its load, it rises in the water (see figure 10). Under normal conditions, the whole dumping process of a split hull barge takes about fifteen to thirty seconds. The rise in barge height is plotted along with its precise position by the black box, giving the exact location of the dump within several feet (which is pretty exact in the ocean).

The black boxes made it easy to track the barges' operations and have a precise record of them. This facilitated site mapping and management, prior to getting actual bathymetric information from scheduled monitoring surveys. This real-time information made it much easier to plan moving the disposal operations to another area to begin a new mound, or to fill in gaps between mounds. Like other electronic equipment, though, the black boxes were also sometimes finicky, and this made them a nuisance to the dredgers. When running under a contract operation with tight time schedules, problems with surveillance devices were just another headache that could cost precious time to fix. It was becoming clear, though, that there was a new era of ocean disposal at hand, and it would require some adjustments from those who wanted to stay in the game.

The environmental groups were a constant presence in the background. Clean Ocean Action was not the only one involved. The Environmental Defense Fund worked with COA, and it came up with some ideas to help solve the developing dredging crisis. One of them was to dig borrow pits in the ocean instead of using the existing ones in the harbor, which had been the source of so much public antagonism. The problem that the agencies had with this idea was that the pits would be in the ocean, and the regulations specifically prohibit the disposal of "unacceptable" material in the ocean, pits or no pits. An alternative to ocean disposal was, after all, what was meant for the more contaminated material; "clean" material was generally assumed always to have a place in the ocean or elsewhere and was not really a problem. Contaminated material would have to be handled differently, an example being borrow pits in the harbor (regulated under the Clean Water Act). The testing revisions had not yet resolved what would be considered "clean," or whether that meant the same thing as "acceptable" under evolving laws such as WRDA (or in ongoing discussions with environmental advocates).

One of the provisions in WRDA 1986 restricted future disposal at the Mud Dump Site to "acceptable" material only. This was somewhat enigmatic to the agency regulators since the regulations already allowed only "acceptable" material. It was not clear if this was supposed to mean some new, as yet undefined, category of acceptable material or what, exactly. The agencies believed that whatever amount of material that language would turn out to represent, it should not in any case have to be subject to special handling methods like contaminated material. Any alternatives to ocean disposal would probably be of limited capacity, and that precious capacity should be reserved only for material that was unacceptable for ocean disposal. The environmental groups did not openly oppose this concept, but their proposal would mean placing contaminated material within the boundaries of the ocean and hence violating the law. The environmental groups were politely informed by EPA that the ocean pit idea was not one that could be considered under the present regulations. They seemed to accept the position then, but it was apparently an idea of great appeal to them since it reappeared a time or two again.

The apparent difference in the way the regulations treat disposal of dredged material in the ocean as opposed to inland waters may seem nonsensical, but it is based on the differences between two environmental laws and actually makes practical sense under some conditions. Ocean dumping is regulated under the Marine Protection, Research, and Sanctuaries Act (MPRSA), while disposal in inland waters is regulated under the Clean Water Act (CWA). The CWA and the regulations promulgated from it are somewhat less specific on the effects to be avoided and the testing required, compared with MPRSA and the Ocean Dumping Regulations. Therefore, there had been no uniform procedures or guidelines set nationally for disposal of dredged materials in inland waters, and the various states and regions applied whatever guidelines they felt were appropriate. The states have more authority under the CWA than in MPRSA, and they can be delegated some of EPA's authority under parts of this law once they have been properly qualified.

The Corps was obviously the major player for inland-water dredging/disposal also, and it had some years before this coordinated with EPA and various states evaluation protocols that were, basically, as liberal as could be agreed to by the relevant states or EPA regions. The Corps would have obviously had no interest in making these more restrictive if it did not have to, and it knew that promulgating changes in so fundamental a law as the CWA would not be a trivial exercise for EPA. So, dredge testing and evaluation methods under the CWA remained a hodgepodge of various protocols around the country. Many relied solely on a few sediment contaminant standards that reflected the background sediment conditions in the ports and harbors to be dredged, or in the areas just outside of them that would be used for disposal.

There were no inland (CWA) areas that were actually using bioassays at this time, as was the case in most coastal areas where ocean dumping was occurring. This was even the case in the Great Lakes, which, with their large number of important ports, one might think would have been regulated in a more serious manner. As the EPA Region 2 dredging program began to look more closely at the ports in Lake Ontario under its jurisdiction, this seeming paradox became apparent to the author and Marine & Wetlands Protection Branch chief Mario Del Vicario during a visit to Corps offices in the region. There were also surprisingly few records maintained on disposal operations by the Corps in some areas, and there was much less monitoring of sites, so several meetings in the Great Lakes area with the Buffalo District Corps and NYSDEC were arranged by Region 2 to promote upgraded testing and site monitoring.

The above-mentioned approach using background sediment values which was generally used in the Great Lakes, in Long Island Sound, and to some degree also for the Port, adopted the strategy of "no further degradation." The principle here was that if the contaminant levels in project sediments were not greater than in areas that were proposed for disposal, then ostensibly the disposal site would not be worse off after dumping.

For years, dredging evaluations in the Port used a similar strategy for evaluation of bioaccumulation test results. Even this was a step beyond the approach being used in fresh waters, however, since the Port program was using measured or calculated *tissue* values that represented the "background." "Matrix" values for four major contaminants, and for each of the two test organisms, had been developed based mainly on average tissue concentrations in marine organisms from the general area of the ocean disposal site. A few of the matrix values were calculated from measured ocean water concentrations of these contaminants in the area, owing to lack of biological field data. The New York regional matrix values were nevertheless an elevated form of comparison with background values, using measured or calculated bioaccumulation tissue concentrations. The fresh water ports compared directly the sediment chemistry of proposed dredging project sediments with that of "reference area" or background sediments.

For the sediment-based no-further-degradation strategy to be environmentally protective, the devil again is in the details. A crucial aspect is the degree to which the data sets for the dredging area and the disposal area are appropriately characterized and compared. By using a less than adequate number of samples to characterize an area, or by compositing unlike areas inappropriately and causing a "dilution" effect, or by using inappropriate statistics, a dredging area that is considerably more contaminated than an outer disposal area can be made to appear not to be. States and EPA inland regions tried to hold the Corps' feet to the fire on these details to the extent they could, but it was a constant battle because there were always economic issues raised when it came to the dredging needs in industrialized harbors. (More recently, the Inland Testing Manual, which closely mimics the Green Book, has been implemented by EPA and the Corps nationally. It acts to conform the CWA to the MPRSA without the need to change the CWA, though it was not being followed very strictly in all parts of the country at the time of this writing.)

The MPRSA, on the other hand, had always required the strict testing and evaluation procedures that are partly discussed in this book. So therein lay the difference in protection afforded by law between ocean and inland waters. From a practical standpoint, however, at least with regard to borrow pits, the differences in the legal framework between ocean and nonocean waters actually make some sense. The operational management of borrow pit disposal (which in the ocean would have to include the pits' initial construction) is much more feasible in the shallow, protected waters of a harbor as opposed to the open ocean. Conducting accurate and safe disposal, cap placement, and monitoring operations on a continuing basis, including in rough weather conditions, would be much more manageable within a protected harbor area. There are two main benefits of conducting borrow pit disposal operations in protected harbor waters: wave energy capable of resuspending sediments is much less of a factor, and the shallower waters make for more accurate and controlled disposal and capping operations.

The Corps had in September 1991 come up with a new proposal for management of dioxin and decided to convene another dioxin steering committee meeting. It had been EPA's position that any dioxin accumulation (at a detection limit of 1 pptr) in test organism tissue would mean the disposal project had to be capped with "clean" material or sand. If dioxin accumulation was greater than 10 parts per trillion, then the project was a Category 3 and the dredged materials could not go into the ocean. The Corps proposed now that up to 4 pptr should be a lower limit (instead of any detected accumulation) below which capping would not be required. It based this limit on the results of the recent chemical monitoring of the Mud Dump, which had found an average of about 4 pptr in various organisms from the surrounding area. The second part of its proposal was the real kicker, however; there would be no upper limit for accumulation as long as the material was capped.

This meant essentially that the Corps believed capping was an effective cure for any level of contamination in material going into the ocean, as long as it passed the toxicity tests. Even if the Corps maintained that bioassay failures would still fail ocean disposal, that line would become blurred if any level of dioxin bioaccumulation could be found acceptable with capping. None of the other agencies represented had that level of confidence in the ability to cap material in the open ocean. To have no upper limit for test organism accumulation meant you could have even the most highly contaminated sediments in the Port, based on bioaccumulation, being qualified for the ocean. Though highly contaminated sediments would be expected to fail the toxicity tests, it would be possible for this not to occur. For example, contamination levels high enough to cause serious food chain effects may not be acutely toxic to some test organisms.

Nevertheless, a couple of months later, the Corps surprised everyone by sending out an officious letter to EPA and the rest of the dioxin Federal-Interagency Steering Committee, saying it was "establishing an interim guideline for dioxin" based on the "inability of EPA to develop a dioxin guideline." The fact that EPA had proposed a guideline that the Corps happened to not agree with was apparently of little consequence to the Corps. Neither EPA nor the rest of the committee appreciated this move by the Corps. Its interim guideline was not adopted by the committee, and no materials were disposed of under that guideline.

EPA, however, had its own dioxin peccadilloes to follow shortly. By the beginning of 1992 Region 2 had worked out a strict management and monitoring plan for dioxin with the Corps' New York District. Region 2 management had later that year unhappily decided (under intense lobbying by the PANYNJ and other dredgers, and pressure from the Corps) that with strict capping management and monitoring requirements, materials testing at a tissue level up to 25 pptr could indeed be allowed into the ocean. The decision was that very strict and protective measures for capping could be required that would effectively minimize any potential for exposure. The reasoning also was that there were not sufficient scientific data to require a 10 pptr cutoff level, and that the newly issued Green Book had (incorrectly) included the FDA level of 25 pptr. Before the end of that year, meetings had been held with the environmental groups, and the above reasoning was discussed with them. The groups did not like this turn of events and did not buy the reasoning.

Then, following more internal discussions, EPA headquarters determined that the FDA dioxin 25 pptr level should not have been adopted in the Green Book and was a mistake, and by the end of the year Region 2 sent another letter to the New York District setting the cutoff level back to 10 pptr. It was now the Corps' turn to be upset, and it wasted no time in voicing its concerns over this turn of events. EPA's reasoning to the Corps was that there was not scientific evidence to support the 25 pptr level. In addition, EPA said that the Green Book inclusion of the 25 pptr FDA number was wrong, that there was considerable concern among the environmental community about the more lenient approach, and that capping could be appropriate only for levels up to ten times the unrestricted (1 pptr) level. What was unsaid was that there had not occurred any change in the relevant science but only, perhaps, a new way of understanding it. Apart from there having been nothing necessarily wrong with that, EPA's seemingly abrupt return to its former position could nevertheless have seemed somewhat arbitrary and capricious to someone at the business end of the decision.

The "ten times" factor used to set the cutoff value may in fact have been somewhat arbitrary, though it had been applied in other programs as a safety factor when more precise information was not available. It has been used, for example, to estimate a chronic effect value from an acute effect value. Chronic effects are typically those, such as reduced growth or reproduction, that can appear at lower doses than those necessary to cause acute effects, such as death. Nevertheless, the reader may ask, didn't EPA have the authority to set the criteria levels for ocean disposal according to MPRSA and the Ocean Dumping Regulations? The answer is ostensibly yes, but it must be remembered that the regulations retained for the Corps considerable say when it came to dredged material, even the ocean disposal of it.

The established process for resolving regional arguments between the agencies at a higher level requires that the Corps demonstrate why a disposal of materials at a certain ocean site could be supported. EPA then has to provide supporting documentation to dispute those findings. Once a dispute rises past the regional level, the administrator (of EPA) eventually has the final determination. As was noted earlier, however, EPA regional managers are cautious in starting this ball rolling if they are not quite confident in the science behind their position, since they would be called on to refute the Corps' submittals. They would much rather try to work out a compromise with the District Corps than to elevate the dispute to headquarters, especially since the science was (and is) often difficult and subjective.

The most important of all the dredging projects that the Corps had been holding in abeyance pending clarification of the testing was Port Newark and Elizabeth (PN/E) Terminal. The PN/E facility (actually two adjacent jumbo terminals) is the heart of the Newark Bay port system, and is the busiest and highest-volume region of the Port of New York and New Jersey. It is the largest container-handling facility on the U.S. East Coast and was the model for supercontainer ports around the world. (Other ports around the world, including Singapore, are now more highly developed and can handle cargo more efficiently than PN/E.) The PANYNJ owns this property and leases slips to a number of large shipping companies. By the early 1990s, some of the slips had already been out of use because of shoaling and the need to dredge them. If PN/E was not dredged soon, it could be the beginning of the end for the Port. There had already been rumors that some shipping companies were talking about moving to another East Coast port.

Being located in Newark Bay, of course, meant that Port Newark/Elizabeth was also probably one of the more contaminated projects, not only with dioxin, potentially, but also a slew of other contaminants that the area was known to be polluted with. The project had been divided into four "reaches," or sections, for testing purposes, with about ten to fifteen samples taken from each reach composited together for bioassays (chemical analyses for dioxin were done on the individual sediment samples, but samples were composited for bioassays to reduce the high costs of testing). Testing was initiated, and word came that one of the reaches in the project area had failed the amphipod test. Recall that a failure of this (toxicity) test meant no ocean disposal. Dioxin was also found in all four reaches, but at less than 10 pptr in bioaccumulation tissue (just barely less, by the prescribed statistics, in one reach). Therefore, under the agreed-on dioxin management plan, the material from the other three reaches could go to the ocean with capping.

There were no alternatives to ocean dumping that were anywhere near being available. The most likely alternative, borrow pits within the Port, was still far from being ready to implement. All the proposed borrow pit locations were within New York State, and the NYSDEC had asked for additional information on a number of technical details about the Corps' proposed methods. The Corps had not thus far succeeded in satisfying DEC's questions and concerns. The public controversy regarding this approach also had not dissipated at all. The Corps therefore proposed that the reach that had failed be tested again to verify the initial results.

Now, the question of retesting sediments that fail bioassays is a tricky issue from a policy and a technical standpoint. The short answer is usually no, so long as the original tests are deemed to have been conducted correctly. The rea-

Another way of looking at the Narragansett data made a good argument for the 10 pptr guideline as opposed to a significantly higher one (25 pptr). The sediments used for the study were among the most highly contaminated (600 pptr 2,3,7,8-TCDD dioxin) surficial sediments found thus far in the Port, having concentrations that were widely considered by the scientific community to be too high for ocean disposal. Although there was an unofficial soil cleanup level being used in the EPA Superfund program of 1,000 pptr for dioxin, 600 pptr was still widely considered too high for aquatic disposal, because of the more exposed and direct food chains in aquatic and marine systems as opposed to those on land. The Corps would probably not have proposed that these sediments be allowed for ocean disposal. Yet, from these high-dioxin sediment concentrations, the twenty-eight-day bioaccumulation in the Narragansett study was about 19 pptr. Therefore, to allow any material with bioaccumulation under 25 pptr to go unrestricted into the ocean would have meant that even the most highly contaminated sediments in the Port could qualify for the ocean.

Certainly the best way of assessing risks from dioxin in sediments was through finding accumulated tissue concentrations that would result in an effect on organisms. The discussion above is another way of looking at the problem from the standpoint of sediment concentrations and what was the common understanding from that aspect.

sons for this are several. First, bioassays rely on living organisms, which tend to be somewhat variable in their response to a media of mixed toxins, which is usually the case for sediments in industrial ports. The developmental tests for amphipods showed that they should respond fairly consistently to a given sample within a range of about 20 percent. Recall that to fail the amphipod test, the project mortality results had to be statistically greater than those of the reference, and 20 percent greater as well (based on the developmental results). In a close failure (which this project reach was not) the variability can certainly make a difference, so a retest can conceivably result in a pass. The question then becomes, which result is the right one?

Secondly, if enough sediment is not initially collected for more than one round of tests (or if the required maximum "holding times" had expired), the project would have to be resampled. This introduces a whole new range of variability, since contamination in sediments is usually very patchy. Substantial differences in contaminant levels can sometimes be found in samples taken virtually right next to each other. This is because sedimentation is dependent on the suspended and dissolved loads of the overlying water and on the currents and tidal influences (including salinity, temperature, and density patterns in estuarine waters) that govern their movement. This effect on variability is then magnified by the fact that the volume of sediment actually chemically analyzed from a sample is quite small, so a little more amount of a contaminant in a subsample can result in substantial variations on a larger scale.

Sampling requirements are designed so as to try to minimize these variables, but they cannot be eliminated, because that would require exhaustive and prohibitively expensive sampling and testing. Although amphipod tests are not very costly compared with a suite of sediment chemistry or bioaccumulation tests, to require statistically powerful sampling and testing generally would result in ghastly high costs. Sampling and testing plans for projects are therefore designed to be as representative as possible of the project area while keeping costs within some reasonable bounds. Given these variables, it is easy to see that allowing a retest is problematic. If the second test passes, then what? Should a third round of tests be required to make it the best two out of three? It can quickly get ridiculous.

Despite all these problems, retests were still sometimes allowed. The main reason for this was that EPA was loath to be perceived as being unfair to applicants and was therefore prone to go the extra mile for the appearance of fairness. In this case EPA did not authorize retesting, and material from the reach that had failed the amphipod test was not disposed of in the ocean.

In discussions with the Corps on testing for Port Newark/Elizabeth, EPA felt it would be appropriate to use the opportunity to begin phasing in more of the new procedures. The Corps' response was that the old procedures should be used until there was agreement on how to implement *all* the revised procedures. It was taking a very broad view of the language that was in the implementation memo that both headquarters had issued jointly with the Green Book in June 1991. The language in the memo was somewhat ambiguous regarding regional implementation, and the New York District's interpretation allowed the Corps to delay implementation to the maximum. Region 2 had some discussions with EPA headquarters to try to clarify the memo writers' intent, but these did not result in speeding the time frame very much. This would be standard operating procedure that the New York District would use for some time. As long as there was no agreement that dotted the i's and crossed the t's (and the Corps would find a good many of these), no additional test updating would be required for applicants, even if the new requirements appeared to be of obvious merit to Region 2. Amphipods and dioxin were the only new items in the revised Green Book that were being implemented (though many in the dredging community were probably thinking that these two alone were already more than sufficient).

One of the founding principles of the Green Book is the concept of tiered testing. It is based on the requirement for doing only the amount of testing adequate to make a supportable regulatory decision, no more and no less. The tiers go from tier one to tier four, from the least amount to progressively more exhaustive and expensive tests and evaluations. In tier one, any relevant available information on the dredge site or material is collected and evaluated, including previous sediment evaluations, information on any spills in the area or other likely contaminant sources, and previous dredging and disposal activity. Sometimes a decision can be made at this level, especially if the project is in an area that is unlikely to have significant contaminant sources and/or is in a high-energy environment such as an inlet with fast currents, sandy sediments, and little likelihood of fine-sediment deposition.

Tier two involves the application of mixing models and bioaccumulation models (both discussed later). These estimate the potential effects generally of disposing of the material, and effects on the ocean water ecology in the case of the mixing model and on the benthic-based food chain in the case of the bioaccumulation model. For both of these evaluations sediment chemistry is necessary, along with other data on oceanic conditions in the case of the mixing model and with other food chain information for the bioaccumulation model. The results of the mixing model are compared against marine water quality criteria (explained in more detail later). The bioaccumulation model is fairly conservative; that is, it tends to slightly overestimate a potential to bioaccumulate, from a project sediment's chemistry results. So project proponents usually will opt to go directly to tier three bioaccumulation testing.

The bioassay testing included in tier three encompasses water and sediment-based toxicity bioassays, and sediment-based bioaccumulation tests. In some parts of the country, decisions are sometimes made at one of the lower tiers, if warranted by the available information and data. In the Port, because of the known contaminated nature of the sediments, almost all decisions have to be based on tier three bioassay testing results. There had been no disagreement on this (the relative ease of passing the earlier bioassays may be a factor here). The only projects with sediments that went into the ocean (or used for beach nourishment) based on tier one evaluation had been several reaches of Ambrose Channel, at the harbor' mouth, whose sandy sediments and high-current profile ensured that little if any concern for contamination existed. Tier two evaluation was not even necessary for these sediments, as provided for in the regulations for these kinds of areas.

Tier four is reserved for special cases in which regional decisions cannot be made even from the results of the tier three bioassays, for reasons involving the complexity of the findings or the interpretation of them. Tier four evaluation may involve long-term studies or other more critical testing than is routinely conducted in tier three. These procedures are rarely, if ever, deemed to be necessary. The long-term dioxin bioaccumulation study conducted by EPA Narragansett Lab discussed earlier can in a sense be considered a tier four evaluation, since it investigated longer-term exposures to help the interpretation of standard twenty-eight-day test results.

An example of a testing change that almost every other region was implementing was to use the whole sediment for the solid-phase bioassay tests. In the old Green Book, the sample for these tests came from the process that was used to make suspended-phase test samples (the water-based toxicity test). The procedure called for mixing one part of the sample sediment with three parts of seawater and allowing time to settle, all under specified conditions and time periods. The settled part was used for the solid-phase toxicity tests, and the rest of the sample that stayed in suspension was used . . . in the suspended-phase tests. These tests are done with organisms that normally live in the water column and that are usually different from the benthic organisms used for the solid phase. It had been widely accepted by then, however, that for solid-phase tests the sample as collected at the proposed dredging site should be used directly, without mixing with water as in the old methods. This is because the mixing could potentially dilute out contaminants in the sample prior to solid-phase test exposures. Even this change the Corps would not agree to implement until a completed manual was agreed on. It was EPA's view that this was not scientifically appropriate, but at this point in time the New York District Corps still held more sway over the dredging program in general and the implementation of new tests than did Region 2 EPA.

The year 1992 began with continuing disagreement between EPA and the Corps on the implementation of Green Book revisions. To complicate matters on the EPA side, reshuffling of management occurred a couple of times between programs in the Water Division, of which dredging was one. This was standard procedure in EPA to give managers a broader experience over a number of different programs. It may be an appropriate management tool in many cases, but the dredging program at this time required a consistent focus and approach. There was certainly no such shifting on the Corps side; its managers were consistently focused. Following one of these shifts at EPA in January 1992, a "fresh" negotiating approach undoubtedly resulted in unnecessary delay in reaching agreement on key issues. These shifts resulted in backsliding from movement that had been going in the right direction, and in all cases the fresh starters were disappointed after several negotiating sessions with the Corps and soon got back on course. That course was to recognize the irresistible factors that were on EPA's side in the long run: the Green Bank revisions were scientifically the right thing, there was a fairly wide scientific consensus on that, and the public wanted these changes (at least the environmentally involved public).

As it was becoming clear that the environmental groups were staging for a lawsuit on Port Newark/Elizabeth, the PANYNJ decided to throw a new plate on the table. It had hired a consultant to do a risk assessment on the potential effects of disposal of PN/E material in the ocean. Risk assessments had become popular in many environmental programs as a way of trying to better gauge potential adverse effects of an activity while taking into consideration as many of the real-world factors as possible. Risk assessments can be responsive to concerns of industry and also be environmentally protective when done right. Industry generally likes risk assessment because it gives them a chance to put something on the table first with a scientific imprint on it. Regulatory agencies use the techniques to better define what might be the actual risks to the environment through a process with transparency for the public.

The key elements of risk assessment are to accurately describe and quantify the toxicity of the offensive source, identify the specific resources such as populations of a particular fish, bird, or mammal (including humans) that are threatened by the activity, and understand the pathways by which the toxicity can affect the resources. In a formal risk assessment quantified values are assigned to each of these elements in a set of mathematical equations, and the results are combined to arrive at a calculation of risk for each identified resource. The exercise may have to be repeated if there is more than one likely exposure pathway, and likewise if more than one contaminant is being considered. The risk can be in the form of carcinogenic risk, such as a one in a million increased risk of cancer, or some other kind of malady to animals or humans, depending on the kind of toxicity associated with the source contaminant(s).

A risk assessment can be an accurate portrayal of the actual environmental risk posed by an activity or a situation. Its accuracy depends greatly on a number of different factors. One is the availability and soundness of information on each of these elements, and how well the job is done of actually identifying and obtaining the relevant information. It often occurs that estimates are made for elements that can be better quantified by more diligent searching or even, if necessary, by conducting some level of scientific work such as a field survey or lab work. This sort of attention to the quality of information will dictate how accurately the elements are quantified at each information point. Lastly, but really underlying the whole process, is the truth of the assumptions on which the overall approach is based. So risk assessments can also be quite off base, especially if the underlying assumptions are not accurate. A lesson in risk assessment procedures can be obtained by inspecting the PANYNJ risk assessment for the Port Newark/Elizabeth project.

The PANYNJ risk assessment did a thorough job of describing and quantifying many of the elements listed above, but its underlying assumption was faulty. It assumed that the main route of contamination to fish is the dispersion through water of dioxin from sediments. Though it correctly presumed that fish are the final link to humans, the major pathway to fish contamination from sediments is not through water but is instead primarily through the benthic food chain. The true exposure route is from sediments directly to benthic invertebrates such as worms, small crustaceans, and clams, then to small fish, then to bigger fish. This pathway is direct and is much more conservative; that is, more of the sediment contamination will potentially reach the fish people eat. The sediment-to-water-to-fish pathway is inherently much less conservative because it is much less direct—only very small amounts of most contaminants dissolve from sediments into the overlying water, and the dissolved amounts are then widely dispersed. In this scenario fish would obtain contaminants only through gill uptake from water that contained very diluted concentrations in comparison with any sediment concentrations.

Therefore the mathematical model for the water-uptake scenario uses much smaller transfer factors for contaminants moving between the three media, and thus results in much smaller estimated concentrations in fish.

The PANYNJ's finding of a very small additional risk from the Port Newark/ Elizabeth sediments was therefore not believed by EPA scientists, including the regional risk assessment experts. After meeting with EPA, the consultant went back and had another go at it, but this was also not acceptable, based on some other technical but important factors. After EPA reviewed the second version, it sent a letter with technical comments and offering to work with the consultant to help address them, but nothing further was heard from the PANYNJ on its risk assessment.

The risk assessment had not fully evaluated all the potential pathways for the source of the risk (contaminants in dredged sediments dumped into the ocean) to reach and affect the potential end receptors (fish, wildlife, or humans). The PANYNJ assessment had missed not only a potential pathway but actually the most likely pathway. A truly comprehensive risk assessment would have not only assessed the benthic pathway but also modeled more than one food chain in this pathway. Separate food chains could include the contaminant routes to higher-level fish, such as bluefish or tuna, and to lobsters, which may have different uptake potentials of contaminants from benthic organisms and from sediments than do fish. For example, bluefish and tuna feed less directly from the benthic environment than do lobsters or even other bottom-oriented fish such as flounder. Bluefish and tuna are one or two steps up the food chain from more direct benthic feeders that make up part of their prey, which can include cunner, hake, and other small bottom-oriented species. And these consumers of benthic prey may constitute only a minor part of their overall food sources. Tuna and bluefish are pelagic (open water, free-swimming) species that feed mostly on schooling pelagic prey such as anchovies and herring. Finally, they feed over a larger area, typically, than lobsters and bottom fish and would therefore be assumed to have less of their overall food intake from organisms affected by an ocean disposal operation.

On the other side of the equation is that bluefish, for example, have very high feeding rates and typically consume more mass per body weight than their bottom-living counterparts. Therefore, the assessment should include an estimate of the percentage of time that might actually be spent feeding in the area of the disposal site as well as estimates for feeding rates. This kind of information may be gotten from tagging and other field studies. Another important information element would be the portion of their food sources that are made up of bottom-oriented prey species. Information of this type can be obtained from stomach analysis studies.

Another factor in this pathway when it is extended to potential uptake in humans is the seafood consumption practices of fishing populations living in the vicinity of the disposal site, or of other populations that may consume seafood affected by the disposal activities. This risk factor itself includes many variables. These range from what species the target populations typically eat to what portions of the organism they typically eat, how often they eat seafood (from a potentially affected area), and how the seafood meals are prepared. Most humans eat only selected types of seafood among all the species that may be affected by ocean disposal. Most humans also eat only certain portions of these species. So, taking for example our bluefish, a high-fish-consuming population of recreational fishers living along the New Jersey shore can be estimated to eat a certain number of meals of bluefish from the vicinity of the Mud Dump Site. This can be estimated (and has been in subsequent risk assessments) from fisher surveys conducted by NJDEP and the National Marine Fisheries Service.

Other factors that should be included in the assessment (using mathematical estimates) are a consideration that only the muscle and not the organs of the fish will be consumed and, as well, the lipid content of the muscle. The lipid (or fat) content is all-important as it relates to the likelihood of contaminant uptake—the higher the lipid content, the higher the likelihood. Organic-based contaminants such as PCB or dioxin will be preferentially accumulated in the lipids of fish and other organisms. The method of meal preparation can also be a factor in the projected uptake of contaminants by a human population. A cooking technique that allows for the release and discarding of fat, such as broiling or barbecuing, can significantly reduce the potential for uptake of these contaminants. (These are in fact measures that are often recommended in state fish advisories.)

Going back to our lobster pathway, the potential uptake of contaminants from benthic prey is more direct, since lobsters live on the bottom and feed directly on worms, clams, and other crustaceans that live in or on the bottom sediments. Therefore there is a more direct link; lobsters are the next step in the food chain. Lobsters may also spend longer times, compared with pelagic fish, feeding within the disposal site or the immediately surrounding area. By all these exposure factors, lobsters may be estimated to potentially accumulate more contaminants (measured as being in higher concentrations in their tissues) than bluefish. But another important factor mentioned above may offset that estimate. Lobster muscle has much lower lipid content than bluefish.

So, for humans that eat only the lobster muscle, there may be less risk than for humans eating bluefish.

The above example is only a limited view of how one part of a risk assessment is done, but it should elucidate how critical are the appropriate identification and quantification of the various elements. All the relevant factors and pathways have to be estimated and understood as inclusively and accurately as possible, since elements within some can cancel out others or enhance the resulting estimate. Bluefish may be less directly exposed to sediment contamination than lobsters because of their food chain, which would decrease their estimated uptake, but bluefish have higher overall feeding rates and higher lipid contents, which would tend to increase the estimates. Lobsters may have a more direct food chain exposure, but their lower feeding rates and lower lipid content (in muscle) would tend to offset that factor in a risk assessment.

In the meantime, the Corps was starting to soften on some of the key testing revision issues. Perhaps it *could* provide some additional information to EPA when a permit application was received, so that the right sampling and testing could be coordinated. And maybe it could look a little more closely at including the requirement to do chemical analysis of project sediments. Recent articles in the (Newark) Star-Ledger had brought the dredging crisis home to a wider audience in New Jersey. There was concern expressed for the ocean environment, but there was also concern expressed by some New Jersey politicians about the threat to local shipping interests. The environmental groups kept asking about the status of implementation of the modified tests. The writing was on the wall, and it appeared that the Corps was beginning to pay heed. A revised regional testing manual was issued to the testing labs in June 1992 to get their take on any potential technical problems with the revised methods. This early version did not have all the issues reconciled, but it included most of the revisions that were being implemented or considered for implementation.

There occurred during this time an interesting sidelight to the dredging program that Region 2 was involved in having to do with ocean research. The EPA coastal regions receive annual funding from headquarters to carry out their ocean dumping programs, and they have fairly broad leeway in how the funds are expended, as long as certain baseline functions are conducted. (Cutbacks in funding for EPA in recent years have significantly reduced these discretionary funds.) One of the more unusual projects that was funded by Region 2 was research at the Narragansett lab relating to the dolphin and seal die-offs that were occurring along the East Coast. The study was interesting to the regional office in the broader context of evaluating all aspects of coastal ecology that could have relevance for dredge disposal decisions. The ecology of the coastal ocean, as in most environments on the planet, includes many interrelating components that are not always well understood. These components can be "physical/chemical," such as variations in water chemistry, temperature, and movement, or biological, such as species interrelationships and their interactions with the physical and chemical world. It therefore makes sense to try to understand as much as possible these kinds of relationships, which may help in making appropriate regulatory assessments for ocean disposal.

The study involved DNA and chemical analysis of tissues from beached sea mammals. This kind of data, along with data on other marine organism tissue and water chemistry, can be applied in evaluating the possible relationships between causes of death or disorientation in these animals and environmental contaminants. It had been postulated that contaminants (which had previously been found in varying concentrations in the fat of these animals) could be causing immunosuppression, which could weaken the natural resistance of whales and dolphins to disease. This kind of information, if verified and quantified through food chain work, could theoretically be applied in improving risk-based approaches for evaluating dredge test results, such as the bioaccumulation tests.

There were, however, no definitive indications in this regard from the analyses at the level of study that was done. The study did, however, add to the growing amount of information on this science. Human activities are being found to affect even the most remote ocean regions, whether by chemical contamination through air or water or by way of floatables and other solid waste pollution from ships and shore. Ocean disposal is a visible and easily identified source of potential pollution to the marine ecosystem, and it will always be under scrutiny—as it should. Monitoring and funding studies like the one involving dolphins and seals is therefore relevant in managing a dredged material ocean disposal program.

The environmental groups did not tip their hand regarding a lawsuit until that summer. A Freedom of Information Act (FOIA) request was received from the Environmental Defense Fund for all documents having to do with development of a dioxin strategy. (The Corps and EPA had made public their development of a dioxin monitoring and management plan, which incorporated the threshold numbers.) Materials from the Port Newark/Elizabeth project would be disposed of at the Mud Dump Site under the plan's strict requirements. The ball had started rolling toward a lawsuit. By the time EPA lawyers had completed their review of the requested files it was in late November 1992, which is when they were sent to EDF.

There quickly followed a meeting between the Corps and EPA; Clean Ocean Action, Environmental Defense Fund, and other environmental group representatives; and representatives of the state environmental agencies from New York and New Jersey. The Corps talked about the forty-project backlog that had built up and described the plan to address it. The plan was for one year of disposal for the projects that passed the tests currently in place, followed by six months of monitoring afterward. EDF and the other environmental groups wanted to know how the federal agencies would be able to mitigate the effects of the Port Newark/Elizabeth disposal if the effects would not, in reality, be known until after the dumping was completed and the monitoring done? The response was that the federal scientists had estimated the potential effects to be negligible, based on the extensive testing and evaluation that had been done.

The groups were not convinced, citing that, for one thing, not all exposure pathways had been adequately evaluated. Clean Ocean Action wanted to know whether EPA had considered and evaluated the fact that the dump site was a popular fishing area and that large schools of bluefish could be drawn to the disturbance of a dump event and feed on the contaminated organisms in the dredged materials. This was promised to be further looked into. A researcher from Ramapo College who had been conducting field studies on dioxins in the Newark Bay and Kills area, the general location of the project, described her findings of high levels of dioxin in crabs, clams, and worms collected there.

The environmental groups asked why EPA had gone from 10 to 25 pptr in its tissue thresholds for dioxin, and hinted at the possibility of political motives. The Corps graciously stepped up for EPA and expressed mild shock at the suggestion. EPA noted the apparent lack of scientific support for the 10 pptr level. An EDF scientist noted that some preliminary results of the EPA national dioxin reassessment indicated that even very low levels of dioxin should be of concern. This was duly noted, but it was indicated that coordination with the EPA headquarters office responsible for the reassessment had occurred, and the regional plan was more or less sanctioned. Clean Ocean Action asked whether the revised testing procedures had been used for the project. The

Corps gave its standard response; there were no criteria on how to evaluate some of the new tests, so they could not reasonably be required. The reception of the agencies' plan for addressing the backlog was less than lukewarm. The meeting ended with a sense that the environmental groups were very dissatisfied with the state of affairs and that, if a goal of the meeting had been to head off a lawsuit, that goal was probably not achieved.

The last month of 1992 is notable for two important events. First, the revised regional testing manual, the *NYD Corps of Engineers/EPA Region 2 Guidance for Performing Tests on Dredged Material Proposed for Ocean Disposal,* was issued on December 18, 1992. For many in the dredging community, this date will always be associated with implementing the testing changes that cut off ocean disposal for many dredging projects, thereby precipitating the dredging crisis in the Port of New York. Of course, the crisis had effectively begun earlier, when the Corps started to hold up its processing of projects for dioxin testing and, soon after, for the amphipod tests. The testing changes were certainly the trigger for the crisis. The powder charge in the bullet, however, was more certainly that there were no alternatives to ocean disposal that had been developed and in place to address this readily foreseeable eventuality.

The issued testing manual included the dioxin management and monitoring plan as a key attachment, and it clarified the testing requirements for all new projects. Unfortunately, it did not clarify the requirements for projects that the Corps had held back; some of these projects' applicants had already done some of this testing. This issue of "grandfathered" projects was to be a matter of continuing contention for some time.

The other important event in this last month of 1992 was a big storm, a heavy nor'easter that shook the New York/New Jersey coast and the Mud Dump Site. Bathymetric surveys that were done after the storm indicated that large amounts of sediments at the site had moved. The scientific/engineering consulting firm that had been doing these surveys for the Corps evaluated the data and in March 1993 completed a sobering report on the conditions following the storm. The survey had been done in February 1993. The results from this were compared with those from a previous survey that was done in November 1992. The results were quite staggering. There was an unaccounted loss of 80,000 cubic yards of material from certain sections of the Mud Dump Site that had been there in November. Also, much of a projected 200,000-cubic-yard mound that should have been in the southeast quadrant of the Mud Dump Site from recent disposals since November '92 was not there. Part of the reason why material was missing from the southeast mound could conceivably be sloppy dumping practices; much of the material that was supposed to be dumped there may not have ended up there. It was more difficult, however, to provide a plausible explanation for the loss of the preexisting 80,000 cubic yards of material that had been located at the site before the storm. The consultant concluded that, essentially, only material on seafloor deeper than about sixty-five to seventy feet had not moved.

This information was quite unsettling to the agencies. EPA now viewed some of the Corps' prior statements about site stability with newfound skepticism. The Corps looked to its Waterways Experiment Station to review all the information and hopefully extricate it from this mess. These findings could torpedo the agencies' plans for managing dredged material in the region, which had been outlined in the recent public meeting as including a period of disposal under the existing requirements followed by monitoring. Borrow pits were certainly not ready for prime time. Even though the Corps had gotten a strong thumbs-down on borrow pits at the public meeting in Brooklyn, it had not given up. It maintained it had given the NYSDEC all the information required to approve its request to dispose of dredged materials in pits within New York waters but that DEC had not been responsive.

Apart from needing NYSDEC's approval for the project to satisfy regulatory requirements, such approval would also have provided needed support to help the Corps resuscitate its borrow pit alternative from the public's perspective. Although the public may not have perceived NYSDEC as having a pure white hat, DEC nevertheless was viewed as more the good guys than the Corps was, and its support would have added weight to the Corps' position on borrow pits. This doubtless added to the Corps' ire against the state. For its part, NYSDEC said that the additional information it had requested and questions it had raised in a number of technical areas regarding the Corps proposal had still not been adequately addressed.

EPA had lashed itself to the Corps' suite of alternatives, since it did not have any others itself and could not at this point turn a blind eye to the situation. So disposal with capping at the Mud Dump Site was the only ticket for the foreseeable future, at least the next year or so. Yes, there were plans for developing other near-term alternatives such as land-based disposal, constructed underwater pits (as opposed to borrow pits), even sediment decontamination. These were still only conceptual, however, with nothing really on the drawing board for implementation in the near term.

The impact that the poststorm report could have on the planned, upcoming disposals was therefore potentially devastating. If so much material could have moved after only one major storm, what did that say about the legitimacy of capping questionable materials in the ocean? Worse yet, what did it say about the overall strategy of a containment site in that location, about the ability to place material in that area of the ocean and expect it to remain there? These were questions that the agencies sincerely hoped would be favorably addressed by the anxiously awaited assessment from WES.

At around the beginning of 1993, the regional EPA and the District Corps were also dealing with a somewhat less troubling problem: how to spend $2.7 million. The Water Resources Development Act of 1992 had allocated that amount of funding to EPA for identification and development of decontamination technologies for New York Harbor dredged materials. Although having to spend this amount of money might not seem much of a problem from a private individual's standpoint, government agencies have to abide by strict rules and time frames to do it, or risk losing it. Several institutions and groups were investigated that could manage the work, and to which the funds could be obligated. Any projects funded from this allocation would be fiscally and operationally managed through this chosen entity. WES ended up getting the lion's share of the funding to conduct investigations on decontamination technologies, and a small amount of other work was contracted to academic researchers. The important point here is that the federal government in Washington, D.C., had acknowledged the problem in the Port, had allocated substantial funds (on top of previously allocated funds in 1990) to help address the problem in the Port, and from its standpoint the ball was now squarely back in the regional agencies' court.

7

SEE YOU IN COURT

On January 2, 1993, the Corps issued a permit for the PANYNJ to dredge the Port Newark/Elizabeth facility and dispose of the material at the Mud Dump Site, in accordance with the regional dioxin management and monitoring plan. A requirement of this plan was that every detectable square foot of dredged material placed on the sea bottom had to be capped with a full meter of sand. No matter whether it was a twenty-foot-high layer near the middle of the mound or a centimeter on its very fringes, it all had to get covered with a full meter of sand. Capping had been done before in other parts of the country as well as in New York, but never with this required degree of precision and strictness of coverage. Normally, a three-to-one rule of thumb was used; approximately three times as much cap material as the amount of project material was planned for. After the project material was dumped, the cap material was spread over it, with highly variable degrees of precision for both of those operations. There would usually be some monitoring to assess how well the operation went, and there would sometimes be some additional cap placement if necessary. The plan for the PN/E project was believed to be the most strict capping management plan that had ever been required. And it was to be strictly monitored by traditional bathymetric methods as well as with newer technologies recently developed for examining the ocean bottom. It would prove to be a very telling requirement for the PANYNJ to abide by.

Concerned about the storm brewing in the environmental community, EPA regional executives decided to start a "regulation-negotiation" process with the environmental groups and the dredging industry. "Reg-neg," as it was termed, was the hot new management approach being used in some regulatory programs and was being promoted by headquarters. EPA had started on this path in the early 1990s with its Common Sense Initiative followed later by Project XL (U.S. EPA website 2004). These were described as new ways of dealing with the regulated community in a more cooperative and friendly fashion. The setting of environmental regulation before this had for the most part used a command and control approach. In this approach a proposed regulatory requirement is developed by an agency based on scientific findings, is issued for a period of public review, and after consideration of public comments is finally adopted (or modified or not adopted) through another public notification in the Federal Register. (There have been many proposed regulations that have never seen print as a final regulation in any form.)

The new approach of negotiated regulations could be tailored to specific business sectors or even individual companies. It involved assembling stakeholder groups representing both the regulated and the environmental community and mediators to shepherd along the negotiating process. A mediator first interviews the contending parties to identify the main sticking points, then briefs the regulator on possible remedies, and then attempts to mediate the final negotiations. The ground rules are very democratic. The regulator is granted no more say during the process than the stakeholders. A final, mutually agreed-on outcome is hoped to be achieved by consensus through this process. Persons at EPA headquarters that had gone through the process had opined that it was a difficult way to address highly technical and difficult problems, especially when they are associated with a wide disparity of views. If one were to describe the dredging crisis in the Port of New York, those qualifiers would fit quite well.

Nevertheless, the method was the current rage, and regional executives at EPA decided that it would be worth a try. So a mediator-consultant was hired, and meetings were planned. The mediator soon found that the major issues of contention were not few and were not easy to get a handle on. Nevertheless the mediator was persistent in gathering information, setting up positions, and finally scheduling a meeting. The meeting was very long, went over issues and positions that seemed like déjà vu to many of the participants, and got pretty much nowhere. The process endured for approximately that one meeting.

When the PANYNJ submitted its dredging plan in response to the permit conditions, there was a bit of a surprise. Instead of the 200,000 cubic yards that were initially estimated as needing to be dredged and were stipulated in the permit, the amount now was closer to 500,000 cubic yards. An error of this magnitude is not taken lightly in the permitting process generally, and in view of the high profile the PN/E project had assumed, it was a major issue. The sampling and testing that is done for a project is designed to be as representative as possible of the entire project to be dredged. It is generally frowned on if large amounts of material that need to be dredged are not properly sampled and tested. Normally, small differences can be overlooked since estimates can be a little off, and in the time it takes for longer permit reviews, such as this one, additional deposition of sediments will often occur. But this was not a small difference, and both agencies could not believe that the PANYNJ had botched the initial estimates to this degree. A scheme was prepared for additional sampling and testing of the material that had not been represented in the initial testing, and the PANYNJ had no choice but to do it.

Then, an operational meeting was held with the Port Authority and its dredging contractors to go over the restrictive new requirements that would have to be followed. The Corps went over the responsibility of inspectors and the tight navigational requirements that barge captains would have to comply with, including monitoring the condition of the sea before a tug and barge could even depart for the dump site. This had been required because EPA believed that to achieve the required accuracy in dumping, the tugs and their barges had to reduce forward speed to a minimum while approaching the dump zone and then dumping their load. The tug captains had feared they could lose steerage while towing a full barge if the speed was too low, especially in rough sea conditions. Therefore, it was established that tug captains had to ensure that the weather outlook at the beginning of the trip would be conducive to maintaining a three-knot maximum speed during disposal, some four hours later.

In a subsequent meeting, the federal agencies heard the Port Authority's complaints and last-minute misgivings. It lamented the high operational costs of getting enough sand to finish all the required capping—the project would be more costly than it had anticipated. It was clear to everyone, however, including the Port Authority, that the disposal had to be this way or not at all if the PANYNJ wanted to dump these materials into the ocean. The overall sense of gloom and inevitability that had developed about the dredging situation was such that the Port Authority could not even muster the energy to blame EPA for its troubles, which had been its and the Corps' standard outlet for regulatory frustration. Someone questioned, whimsically, why the Port Authority had to go to all this trouble anyway, since it would probably get sued no matter how the dumping was done. An EPA manager grimly noted that the judge would certainly grant an immediate restraining order, on request from the enviros, if this disposal did not go strictly according to the publicly issued management plan.

As mentioned earlier, WRDA 1992 had come with substantial funds, $2.7 million, to continue development of decontamination technologies. That work was generally proceeding, albeit at a relatively slow pace. The prevailing outlook was not good for actually using any of these technologies anytime soon on anything near the scale that was required for the Port's contaminated sediments. This was mainly due to the high costs of the various techniques (which for most were more than $80 a cubic yard) as well as the difficulty in scaling any of them up for the capacities that would be needed. For practical purposes, new methods are developed in small "bench"-scale studies in labs. The next step up before industrial or institutional application is termed pilot scale, in which quantities and equipment sizes are ramped up to see how the process works under more realistic circumstances.

Nevertheless, claims were heard of the potential for decreasing costs as volumes and competition increased, and the technical results that were coming in actually looked pretty good. The efficiency of contaminant reduction was quite good for many of the technologies that were being tested, with some showing greater than 99 percent contaminant removal. There were other problems associated with some of the technologies besides the costs. Although the more costly technologies, usually involving high-temperature combustion methods, actually destroyed contaminants, they were also plagued by NIMBY (not in my backyard) public concerns because of potential air pollution and other side effects. Other technologies based on separation or extraction were also quite efficient but were also costly and resulted in a highly contaminated by-product (though of much smaller volume) that still had to be disposed of.

Lastly, there were stabilization technologies evaluated that do not remove or destroy the contaminant but instead bind it up by forming a cement-like solid. They are thus generally considered less efficient than the thermal or chemical removal technologies by definition. They are much cheaper, however, since they use low-tech, off-the-shelf methods that include addition of cement and other easily obtained bulk materials, and other waste materials such as fly ash from incinerators or coal power plants. When these methods are applied appropriately, as for the construction of a subbase for parking lots or roads, and thus segregated from environmental exposure, they can be very effective.

At around this time, EPA thought it would be a good idea to put the dredging program under the umbrella of the New York Harbor Estuary Program.

The New York/New Jersey estuary was one of the selected national estuaries to qualify for EPA support under its National Estuary Program; the goal was to bring together all the regional stakeholders and develop a comprehensive management plan to restore and protect the harbor estuary. The Comprehensive Conservation Management Plan (CCMP) for the estuary contained a dredging component that outlined goals such as contaminant reduction in harbor sediments, as part of more holistic contaminant tracking and reduction strategies. Since the CCMP was nearing completion, the program was being viewed as having achieved a measure of success. (In actuality, it is the implementation of a CCMP, the actual achievement of the measures proposed, that is the real test of success for these programs.) Perhaps it was thought that some of this program's good vibes might rub off on the dredging program's bad ones.

The regional dredging/ocean disposal program had by now branched into a number of new areas including nonocean disposal alternatives, decontamination technologies, and investigating the reduction of contaminants going into the Port/Harbor. Perhaps management of the dredging program would benefit from the experience gained in the estuary program in comparable areas. It was under the tutelage of the estuary program that the SMIRFs appeared. This was the Sediment Management Issue Resolution Forum, another meeting group of dredging stakeholders that were assembled to try to resolve the outstanding issues. No one really knows what happened to the SMIRFs. They were less enduring than their television cartoon counterparts.

The major contribution of the SMIRFs was to bequeath the last word of their name to the next form of gathering that the agency executives formulated, the New York/New Jersey Harbor Dredged Material Management Forum. It was finally under the auspices of this dredging forum that the two federal and two state (New York and New Jersey) agencies would forge an impressive number of advancements in alternative disposal options, disposal site management, and the evaluation of material proposed for ocean disposal.

Near the end of January 1993, a meeting was held between the federal and state agencies, Clean Ocean Action, the Environmental Defense Fund, and Coastal Alliance. EPA brought up the SMIRF idea, which included scheduled meetings with the environmental groups, and there was discussion of the decontamination work, criteria for ocean disposal, and other issues. The environmental groups wanted to talk about the Port Newark/Elizabeth project that they were concerned about, both on its own merits (or evils) and as a precedent for continuing disposal in the ocean. They wanted to know how all these future actions would address their immediate concerns. They suggested that a more appropriate dioxin tissue threshold was 3 pptr as a cutoff for ocean disposal instead of the 10 pptr value adopted by the agencies. This was based on a value that NYSDEC had derived and adopted for protection of piscivorous (fish-eating) wildlife (Newell, Johnson, and Allen 1987). Recall that the 10 pptr EPA value was based on protection of humans from consumption of fish. There was no consideration, though, of how the NYSDEC 3 pptr value would apply in a way that was relevant to the uptake factors at the dump site and the mitigative effects of capping. A NYSDEC representative who was present did not suggest that the value should be used directly for evaluating bioaccumulation test results without further consideration of those factors.

The concern that had been previously raised of schools of bluefish swimming into a disposal plume and picking off contaminated benthic organisms was rehashed. EPA staff had canvassed fisheries professors at the Marine Sciences Research Center at SUNY Stony Brook and posed the issue to them. Their view was that the fish would probably tend to avoid such a high-turbidity area and that such an effect had not been observed to their knowledge. EPA relayed this at the meeting, and discussed the more probable scenario.

In a typical dredged material disposal event, three or four thousand cubic yards of muddy sediments are released in short order from the bottom of a split hull barge and fall to the bottom rapidly in a process termed by physical oceanographers "convective descent." The mass of material creates a strong downward flow that brings most of the dissociated material (including most small benthic organisms) down with it. Also, the greatest mass of the material is usually dredged from depths below those where benthic animals live, so their numbers in relation to the huge mass of material would typically be small. Nevertheless, there could be some scavenging of food items that might be available after conditions had settled somewhat. EPA noted, though, that it believed that the potential contaminant levels of those organisms and their potential uptake and the effects on their predators or scavengers had already been taken into account in the testing and evaluation process.

When the subject of a mediated resolution process was reiterated by EPA, the reaction of the environmental groups was tepid. Would the groups consider tabling the disposal criteria for Port Newark/Elizabeth and some limited upcoming projects to work together in a serious way toward measures that would be mutually agreeable? Not unexpectedly to many at the meeting, the answer was no. The PN/E project was too important to be set aside. Although

EPA tried to assure the groups that a new and more inclusive situation would be in place and their concerns would be given the full attention due, it was of little help. The environmental groups' dissatisfaction was palpable as the meeting ended.

The dioxin criteria remained a major issue. Other EPA regions were canvassed for any experience that they might have had with dioxin assessment. Two other regions, in Atlanta and Seattle, had to address dioxin in their dredge sediments. The levels of dioxin in their sediments were much lower than in the Port of New York/New Jersey generally and in the Port Newark/Elizabeth project particularly. The amount of dioxin in their sediments measured in the low pptr, and some areas did not even contain detectable amounts of the most toxic congener, 2,3,7,8-TCDD. Both had conducted rudimentary risk assessment–type evaluations and concluded that materials from the projects in question could be disposed of in the ocean with no capping or other mitigation.

At the end of January, a two day public meeting extravaganza under the dredging forum was held at Colombia University. A mélange of presentations, arguments, rebuttals, and exclamations, with no predominant theme, was the story of the days. The PANYNJ gave a full presentation on the importance of Port Newark/Elizabeth to the Port and the importance of the Port to the economy of the New York metropolitan area. The Ramapo College professor who had earlier presented her findings on dioxin in Newark Bay–area shellfish now discussed new crab data showing that the dioxin levels being found were definitely reducing growth rates in crabs. A commercial fisherman stated that Japanese tuna buyers had told him they would not buy tuna from the region, since it was likely they were feeding on bluefish from around the Mud Dump Site and getting contaminated.

A spokesperson from the Environmental Defense Fund suggested that the federal interim plan should include only this one (PN/E) project. This seemed a very reasonable suggestion given the strong objections that had been voiced previously over this project. It was, however, accompanied by a condition that the agencies could not swallow. This was that each barge load of dredged material from this project must be capped as soon as it was disposed of. In their dioxin management plan, the agencies had utilized the belief shared by most benthic ecologists that colonization by organisms and their bioaccumulation of contaminants to levels of concern takes at least a few days and more likely several weeks. On this project as on most, the disposal schedule would be two to four barges per day, which would result in continually disrupting the dis-

By early 1993 word was coming back from the National Bureau of Standards on a laboratory review study it had been conducting. It had found that there was a large degree of scatter in the results for detection of low levels of dioxin among the laboratories taking part in the review. At concentrations around the detection limit of about 1 pptr, there could be little confidence of laboratories' being really able to distinguish between, say, 3 and 4 pptr. This was no shocker, considering these extremely minute, almost ephemeral concentrations. This was another practical reason to base a regulatory limit on a value, in the range of 10 pptr, that could be measured with some degree of confidence and linked to specific management measures. In this sense, it also further supported EPA's concern regarding the practicality of applying the 3 pptr NYSDEC fish protection level previously proposed by the environmental groups.

posal area as well as continual covering of older with newer sediments. This continual disturbance during any ongoing disposal operation would by itself, without capping, minimize the potential for colonization and uptake of contaminants. The management plan conditions specified the need for continuous operations at any particular area and the need for interim cap placement if there were any delays. Capping after each barge load, therefore, besides being an operational nightmare, was not necessary if the above factors were believed to be true, as they were.

There was discussion of the status of decontamination technologies and their costs. The average costs at that time were given as approximately $80 per cubic yard, prohibitively expensive against the old $4-a-cubic-yard cost for ocean disposal. (Of course the Port Newark/Elizabeth project was going to end up costing much more than that because of the tight controls required and enormous amounts of sand needed for capping.) NJDEP brought out that the Diamond Shamrock dioxin site remediation on the Passaic River was entering a new phase. An administrative consent order was being developed that would provide for cleanup of the river itself, in addition to the land site. Sampling of the river sediments had showed that dioxin concentrations in some places increased considerably with depth in the sediments. This meant that any cleanup of the river would be a tremendous effort, if it ever got off the ground.

From these meetings various new tasks had been identified, and workgroups were formed and staffed by the federal and state agencies to work on them. The workgroup tasks included modifying disposal operations, finding new disposal sites, updating testing criteria, and developing decontamination technologies.

An immediate problem, however, was how to deal with the increased volume of Port Newark/Elizabeth sediments. Both recent and prior bathymetric data had to be compared with the initial sampling information and dredging history, berth by berth, considering the original and new estimated volumes of material, to develop an appropriate new sampling/testing scheme. It was decided that only sediment chemistry would be done, in lieu of requiring the complete suite of biological tests again. But to be able to evaluate this kind of data, a formula had to be developed to determine the comparability of the additional sediment samples and volume to the samples and volume of the original tests. The Corps and PANYNJ were there at every step of the way, to challenge EPA's decisions. Of particular annoyance to EPA staff during this time was that one set of chemistry data returned by the PANYNJ was reported to be in dry weight, as per convention, when in fact it was wet weight and showed much lower concentrations of dioxin than it should have. Though this was surely an honest mistake by the PANYNJ, it nevertheless rankled a bit when it took longer than the day or two it should have to clear it up.

In early February 1993, Region 2 invited several federal expert scientists for a conference on the regional dioxin management plan. Research scientists from EPA labs in Narragansett and Duluth came to New York, and others were on conference lines for part of the day. The scientist from Narragansett, Norm Rubinstein, talked about some results from an ongoing study that was being supported by Region 2. The study looked at uptake of dioxins and PCBs from sediments by sea worms and then by lobsters that were fed the worms (Pruell et al. 2000). He said the data were preliminary but it appeared that the lobsters had accumulated dioxin in their hepatopancreas to levels considerably higher than had accumulated in the worms. This was not a big surprise since it was suspected that dioxin could biomagnify in the lipids of some species as it travels up the food chain. Dioxins, PCBs, and many pesticides and other organic contaminants are "lipophilic," meaning they attach to lipids (fats) in organisms. The hepatopancreas is high in lipid content and is the green organ in lobsters known as tomalley that most people don't eat, although it is apparently relished by some hard-core seafood lovers.

A distinguished research scientist from EPA's Duluth lab, Phil Cook, discussed the results from a major study that he was completing on dioxin throughout the food chain in Lake Ontario. As part of this work he had done an analysis of the data from laboratory studies of dioxin's effects on different organisms. His report included a table that predicted a low-risk concentration for protection of health in fish of 50 pptr (EPA 1993) and for protection of piscivorous wildlife of 0.7 pptr. The reader will recall the 3 pptr NYSDEC level for protection of piscivorous wildlife that the environmental groups had proposed at the public meeting in January. The values of 0.7 and 3 pptr. are quite close but, as mentioned earlier, would have been difficult to consistently discern at the effective low end of detection achievable at that time. An important point here is that the calculations resulting in those low levels for wildlife rely on fairly high and sustained feeding rates by the target species (including mink and piscivorous birds) on the fish containing significant levels of dioxin. This is an example of how different risk scenarios and end receptors can result in very different results. For a human health risk assessment, the *effects levels* used in the calculations may be even lower (i.e., based on cancer effects) than those used for wildlife, but the *ingestion rates* would be much smaller (i.e., humans don't eat only fish). So, the wildlife values can be lower than those for human health based on these kinds of considerations, even though we typically value human life more than wildlife.

The 50 pptr value used by Cook for protection of fish was compared with the EPA Region 2 threshold of 10 pptr in test worms. This was not a direct apples-to-apples comparison since, it will be recalled, the 10 pptr level was to be applied for protection of human health. But it was thought that if the value worked to protect fish, too, all the better. The wildlife value was not thought to be applicable to the ocean site, since even piscivorous birds there would feed over large areas unaffected by the disposal site, and this factor would considerably lower their potential exposure. Given that the protocol would restrict disposal of any material that might result in an exceedance of 10 pptr in fish, it would therefore be protective to fish with a safety factor of five. The EPA managers took comfort in this and determined that the regional 10 pptr threshold applied to the worm bioaccumulation test results was still appropriate.

At the end of March, a congressional hearing to look into the regional dredging crisis was held in Washington, D.C., that included Congressmen Pallone, Greene, Saxton, and Hughes. Several of the congressmen were outspoken. One said that the agencies had better be careful not to create another Newark Bay at the Mud Dump Site. Another opined that he would not lose any sleep if the agencies were sued by the environmentalists. They were all concerned about the Vietnam veterans, who were becoming more vocal about the matter. The reader may recall the earlier dioxin settlement with the veterans and their families—and their lawyers. There were groups of vets that were still following all the national dioxin issues closely, and many had written letters to the agencies voicing their concerns, along with the environmental groups, of adopting too liberal a disposal policy for the ocean. The agencies were accused of ignoring the consultant's report on the massive sediment movement after the big storm. A member of the delegation wondered whether ocean dumping should be stopped and the dredged material temporarily stored in barges until a better disposal mechanism could be found. (The agency members present at the meeting visualized the thousands of 4,000-cubic-yard barges floating around the Port somewhere that would be necessary to hold the approximately 4 million cubic yards of material that had historically been dredged from the Port annually.) The atmosphere was not a friendly one for the agencies. The Corps tried to put the best face on things by noting that WES was reviewing the consultant's data and report and that conclusions should therefore not yet be drawn.

A national EPA/Corps meeting was held in Arlington, Virginia, in late April to go over the Region 2 dioxin management plan and review other dioxin assessment methods going on around the country. The EPA Atlanta and Seattle regions' coordinators described their projects and evaluation methods. An EPA research scientist discussed the most recent science on dioxin and recommended consistent methods of calculating theoretical risk to human health and the ecology. EPA and Corps headquarters representatives discussed the planned development of a formal risk assessment approach for dealing with dioxin in dredged material, with the help of a scientific consulting firm that was familiar with EPA risk assessment procedures. A reason given of the need for this effort was that concerns were being raised by charter boat captains around the Port area of a threat to their livelihood by the public's perception of contamination in local fish. Various aspects of the scientific and environmental information that was needed to conduct this kind of risk assessment were discussed. (Eventually, a draft approach was prepared after several reviews and input by the involved regional coordinators, but the manual was never finalized. It fell victim to the Region 2 dredging crisis that soon engulfed EPA headquarters and higher administration offices.)

In May, the Corps called a meeting with EPA, the environmental groups, and others to present WES's findings on the effects of the big storm at the Mud

Dump Site. First, the physical oceanographer who had originally prepared the report that showed mass movement of sediments after the storm made a presentation. He now talked about the range of error that is inherent in these kinds of assessments. An error range of twenty to thirty centimeters (in depth of sediments) could be expected, he noted, and when compounded by a similar error range in the initial comparison survey, it could be easily seen that trying to make accurate assessments from these methods was fraught with peril. In describing this and other potential confounding factors that could have played a role in the process, he laid the groundwork for the agencies to be able to place a reasonable presumption of uncertainty about the real results of the nor'easter.

It must be clarified that this was one of the most respected physical oceanographers in the consulting business, one with whom some of the agency staff had worked on various projects over the years. The man was widely considered to be a particularly careful and knowledgeable scientist. He was not someone prone to making rash scientific conclusions, as he was now in a sense suggesting had been the case in the original report. For those familiar with his work, it was difficult to believe that those error ranges, or any other pertinent factor, would not have been taken into full consideration when he had first assessed the data and prepared the report. The reality was that he worked for a major scientific consultant that received a lot of work from the Corps and EPA.

A Waterways Experiment Station scientist then got up and talked about some geophysical oceanographic concepts relevant to sediment movement. He espoused on things such as the fact that water deforms more readily than sediment and that waves may feel the effect of the bottom but the bottom might not be affected by the energy of the waves. The overall effect of the presentations was to instill more mystery into the already complicated set of physical data, which had been generated by automated instrumentation and shuffled with computer-assisted statistical methods, and to throw a good measure of doubt onto any preexisting notions on the issue. For the environmental groups, it was beyond their knowledge comfort level to attempt to refute the technical presentation. As for the Corps and EPA staff scientists, none had advanced degrees in physical oceanography, and though these new points sounded a little convoluted and too convenient to some, the complex data sets and potentially confounding factors did leave room for interpretation. They were also loath to question the presentation without very good reason because they knew that both agencies' management would be grasping this new interpretation like a lifeline. So the Corps and EPA ended up skating on this issue for the time being, although later management requirements for the ocean site would reflect the storm-related concerns by implementing more restrictive minimum depth requirements for disposal mounds at the ocean site.

In later discussions about the Port Newark/Elizabeth permit, NOAA raised a concern that the Corps' issuance of the permit may have too readily relieved the permittee from any future natural resource damage claims (which NOAA is authorized to administer) in case the dumped material ended up causing harmful effects in the ocean. The Corps offered to remedy this by allocating funds to conduct a costly study that NOAA and the National Marine Fisheries Service had been seeking to carry out, the chemical analysis of tissues from stranded marine mammals. (The EPA Narragansett sea mammal study was mentioned earlier, and NOAA also had good reasons for expanding the knowledge base in this area. It has considerable responsibilities in the management of sea mammals and providing for the protection of the endangered and threatened species.) NOAA/NMFS could not be blamed for wanting to get support for obtaining valuable scientific data, and even better if the support came out of a potentially risky disposal project. The Corps had fairly deep pockets at this juncture and was only too willing to offer to fund this important work. It also would not hurt if the funding served to get the regional decision-makers at NOAA into a more receptive frame of mind regarding the Corps' overall good intentions about ocean disposal generally, and this project in particular.

A second preconstruction briefing was held with the dredgers to go over the once-more-modified operational requirements. Instead of the need for approaching a buoy at slow speed for disposal, the barges would now be required to steam slowly in defined "lanes" within the site while dumping under strict electronic navigational controls. The reader may recall that similar methods were required in the last stages of dumping sewage sludge at the 106-mile site. Also, New Jersey had required water quality monitoring to measure for movement of suspended sediments into areas away from the terminals during the dredging operations. The project was getting more and more complicated, but at least it looked like it would finally go. Or, at least, it would go for some ways. At the briefing, the Corps notified the dredgers it had received word that the United Fishermens Association was planning to assemble a flotilla of fishing boats at the Verrazano Bridge on a day later that week, to block any dredge barges from going past it to the ocean. The Corps had heard that the event might have media coverage between the hours of 4:00 and 8:00 p.m., so its advice was to refrain from transport operations during that time.

On June 1, 1993, Clean Ocean Action and several other groups filed a request with Judge Dickenson R. Debevoise to issue a restraining order on ocean dumping of the Port Newark/Elizabeth material on the grounds that the project did not meet the ocean dumping criteria. The complaint was against the PANYNJ and the federal agencies (the New York District Corps and Region 2 EPA). On June 7, the judge denied the plaintiffs' application, but he also ordered the defendants to establish that the permit had been lawfully issued under the regulations and the law. The defendants could do this by showing that dioxin was present only in "trace" amounts (a term with legal implications in the regulations), or they could apply for an exception or a waiver under the regulations.

The PANYNJ, curiously, was named as the party required to show that the permit had been legally issued, even though it was the permittee, not the permitting authority. It was clear, in any case, that it would be up to the two federal agencies to defend the permit. As codefendants, EPA and the Corps now began to demonstrate a remarkable ability to work together in (almost) perfect harmony toward a common goal. They immediately began to gather information from their respective researchers, documented the science in their testing manual, and prepared affidavits from managers. Such a lawsuit had not been brought since the National Wildlife Federation lawsuit mentioned earlier, but this one could be even more of a threat to the overall program. Some aspects of the lawsuit went to the core of the ocean testing and evaluation program, challenging important portions of the recently revised Green Book. So the agencies' efforts to marshal the pertinent information were carried out as deliberately and diligently as the seriousness of the situation required. All this would demonstrate, it was hoped, that dioxin was in fact present only in "trace" quantities.

There was no serious consideration given to asking for an exception or waiver. EPA, especially, believed that its science and reasoning had been solid in this permit decision (to not object to the Corps' permit issuance) and that it was mainly the EPA regulations that were in question. EPA headquarters felt that requesting a waiver would be an admission that Port Newark/Elizabeth was a special case, not a strict adherence to the law and regulations. EPA believed, in short, that it was right on the issues and wanted the matter to be decided in court. The judge also had asked Clean Ocean Action to show cause why the dumping should stop.

Meantime, the project was being dredged and the dredged materials were being disposed of at the Mud Dump Site. Most everyone concerned thought that, with the high level of government and public scrutiny of this project, the dredgers would be taking extra care to make sure that everything went exactly according to plan. Well, the first disposal barge that went out—the very first one—missed its target zone by five hundred feet and received a violation from the Corps! The excuse that was given put the blame on Greenpeace. It seemed that some of its members had earlier chained themselves to the barge and, in so doing, had somehow damaged sensitive navigation equipment, which made the barge miss the dump zone. Whatever could be said about the barge operator's navigational skills, whoever came up with this excuse deserved something for imagination and ingenuity. Its Keystone Kops qualities aside, the matter was taken quite seriously. With the already high level of public skepticism about the environmental sensitivity of the Corps and the dredging industry, the agencies felt that this was an especially ill-timed and unnecessary mistake. There were probably some very strained discussions between the Corps, the PANYNJ, and the dredging company. After that, all parties seem to have gotten their act together, since by the end of June three-quarters of the project had been dredged and dumped with no further incident of note.

On July 6, 1993, after having reviewed the agencies' and environmental groups' submittals, Judge Debevoise issued his opinion that the defendants could probably establish that dioxin was present only in trace quantities (and therefore the material would meet the ocean dumping criteria). However, he ordered that additional tests be conducted that he believed were required by the regulations. The tests that had been done were those that were required in the Green Book, some from both the old and the new versions. The crux of the issue was that the regulations could be interpreted to require other tests that were not required in either version of the Green Book. There was, in fact, a type of test that was mentioned in the regulations but that was not required in the testing manuals. The agencies believed that there was good scientific basis for this omittance. They believed that the tests included in the Green Book were the minimal but also the most appropriate ones according to the prevailing scientific understanding at the time. The Ocean Dumping Regulations were written, it must be remembered, to evaluate the potential effects of any kind of material that would be proposed for ocean disposal. Although changes had been made to try to align dredged material testing requirements more closely with those governing all other kinds of ocean disposal (as a result of the previous National Wildlife Fund lawsuit), the job had not been completed in the converse. That is, there were parts of the regulations that required dredged material to be tested in ways that were not considered necessary, by the prevailing scientific opinion.

These included the need to test for bioaccumulation in a suspended-phase test. The language in the regulations did indeed appear to require this kind of test, but there were also other passages that gave the agencies broad latitude in how to apply the requirements. The agencies had agreed long before this that the test made no sense for dredged material disposal, and it had not ever been required. It was not a necessary test in the agencies' opinion, because bioaccumulation is a relatively long-term phenomenon, taking at least several days for appreciable accumulations to occur in aquatic organisms. Dredged material disposed of in the ocean typically descends quickly in a mass, and there is usually rapid dispersion of any remaining suspended material through natural mixing and water movement, in a matter of a few hours. Therefore there would not typically be enough time for aquatic organisms to bioaccumulate significant amounts of contaminants from suspended material. At least that was the prevailing wisdom, and that was what the agencies tried to demonstrate to the judge.

It was shown that organisms living or present in the water column, be they planktonic drifters or active species like fish, would be exposed for too short a time for any appreciable bioaccumulation to occur. So, though a suspended-phase bioaccumulation test was included in the regulations, prevailing scientific opinion in both agencies had deemed it an unnecessary test, and it was not included in either the old or the new Green Book. And it was in fact the Green Book, the agencies argued, that represented the tests that were approved by EPA and the Corps, as was stipulated in the Ocean Dumping Regulations.

In addition, other aspects of tests that were described in the regulations but not in the Green Book had to do with the number of species to be tested. Here, the regulations appeared to require using three different benthic (bottom-living) organisms for the solid-phase tests. The agencies argued, again with the support of prevailing scientific opinion, that the one species (for dioxin) or two species that were used provided a sufficient estimate of the potential adverse effects of the material to the marine environment.

While the agencies were preparing the foundations for these arguments,

SCIENCE NOTE: TEST SPECIES

According to the regulations, for the suspended-particulate-phase tests, species that represented a phyto- or zooplankton, crustacean or mollusk, and a fish all had to be tested. In reality, larval forms of crustaceans, mollusks, and fishes are all in the zooplankton realm. So if a species of larval crustacean or mollusk, as well as a fish are tested, then all three categories are represented, even though by only two species. This was what was being done, and it made eminent sense to the agencies. For the benthic-phase tests, the regulations call for at least one species of filter-feeding, deposit-feeding, and burrowing species chosen from among the most sensitive species accepted by EPA to be tested. In reality, there are organisms that are both deposit feeders and burrowers, and the same logic as for the suspended phase was applied to select two species for testing. In both cases, there was a reasonable argument that the regulations unintentionally required redundant testing.

In the case of dioxin, EPA had conducted long-term studies that showed that one species, a sandworm, accumulated more dioxin than two other species that were also evaluated, and this sandworm was therefore chosen as the worst-case, sole test organism. This decision was based on the science and on the fact that the cost of chemical analysis for dioxin was about $1,400 a sample, and this analysis was required for all the replicate and quality assurance tissue samples needed for one test. An individual test, either bioassay or chemistry, for Port dredging projects is done on one composited sediment sample, which can be a compilation of eight to ten (sometimes even more, unfortunately) individual samples taken from one subdivided area of a project. A large project such as PN/E can have four or more composites. Adding this to the cost of doing all the other chemical analyses for sediment chemistry and the other bioaccumulation tests, the twenty-eight-day aquaria exposures for these, and the toxicity tests brought the total cost of testing one composite New York Harbor sample to well over $100,000. It was these considerations and scientific logic that led the agencies to pursue the legal course in justifying the variance between the regulations and their guidance document.

they nevertheless advised the PANYNJ to do the additional testing that the judge had asked for. This, however, was easier said than done. First, the material had already been dumped into the ocean. This was the first time, to anyone's knowledge, that a sample of dredged material had to be retrieved from an ocean dump site for testing to see whether a permit should have been issued for it to be dumped in the first place. On top of this, there was a sand cap overlying the project sediments. Sampling cores had to penetrate the sand layer to get to the dredged material. This of course was not an insurmountable problem, and with the detailed information that was now available from this disposal operation, appropriate locations and depths for sampling were quickly worked out. There were still other technical issues to work out, however, including what kind of dioxin analysis should be done, and how to verify that the material originally sampled in the harbor was characteristically the same as that disposed of and then sampled at the dump site.

This last item was necessary in order to be able to accurately evaluate the new tests and the comparability of the old and the new testing. Remember that a sampling and testing scheme for a dredging project, no matter how well planned and conducted, is still only a representation of the actual characteristics of the mass of sediments in a large project area. So a recheck of the dumped material would help ascertain how representative that initial sampling actually had been and, by those lights, would also help predict the level of confidence that the original and new testing would be comparable.

Regarding the issue of analyzing or not analyzing the dioxin congeners, the question was whether to require analysis of just the primary and most toxic congener, 2,3,7,8-TCDD (which had been the standard practice till then in the Port and elsewhere), or all seventeen congeners, as was becoming more common for dioxin analysis at the time.

It was decided to remain consistent with the previous tests and analyze only for 2,3,7,8-TCDD. (EPA made the decision to start requiring the total congener analysis, or TEQ, the following March for subsequent projects.) To verify that the material initially sampled was essentially the same as what had been disposed of, it was agreed that only sediment chemistry analyses for dioxin should be done for comparison purposes (and not the bioaccumulation tests). The Corps asked whether it should continue the capping operation while the sampling of the underlying dredged material was to take place. The answer was that of course it should, since capping had been the key to finding the disposal acceptable in the first place.

The other problem was identifying the right organisms for the PANYNJ's lab to use in the additional dioxin bioaccumulation tests required by the judge. For the additional solid-phase tests, identifying two other appropriate species to use was not a problem, since a number of species had been identified and tested at one time or another. However, since the bioaccumulation test for the suspended phase had never been required by EPA, there was no list of approved

Congeners are simply compounds that have similar molecular structures and thus similar properties. The chemical descriptor for the most toxic of the dioxin congeners is 2,3,7,8-TCDD. The numbers describe the molecular positions of chlorine atoms, and the letters describe molecular structure: tetra (four) cloro—dibenzo (two benzene rings)—para dioxin (the position of the chlorine atoms and the two oxygen link atoms); thus, 2,3,7,8-tetrachlorodibenzo-*p*-dioxin (in spoken terms, 2,3,7,8 tetrachloro dibenzo para dioxin). The seventeen dioxin congeners commonly tested actually include ten compounds called furans, which are similar to the other congeners in structure except that they contain only one oxygen atom. The seventeen congeners are equal in toxicity to or less toxic than 2,3,7,8-TCDD and can have up to eight chlorine atoms in their molecular structure.

The toxicity of each congener is assigned according to testing that was done and its toxicity is related to the toxicity of 2,3,7,8-TCDD by a decimal fraction, called its toxic equivalency factor (TEF), listed below (developed by the World Health Organization in 1997). For example, the congener closest in structure (and toxicity) to 2,3,7,8-TCDD is 1,2,3,7,8-PnCDD, with five chlorine atoms and assigned a TEF value of 1.

DIOXINS	TOXIC EQUIVALENCY FACTOR (TEF)
2,3,7,8-TCDD	1.0
1,2,3,7,8-PnCDD	1.0
1,2,3,4,7,8-HxCDD	0.1
1,2,3,6,7,8-HxCDD	0.1
1,2,3,7,8,9-HxCDD	0.1
1,2,3,4,6,7,8-HpCDD	0.01
OCDD	0.0001
2,3,7,8-TCDF	0.1
1,2,3,7,8-PnCDF	0.05
2,3,4,7,8-PnCDF	0.5
1,2,3,4,7,8-HxCDF	0.1
1,2,3,6,7,8-HxCDF	0.1
1,2,3,7,8,9-HxCDF	0.1
2,3,4,6,7,8-HxCDF	0.1
1,2,3,4,6,7,8-HpCDF	0.01
1,2,3,4,7,8,9-HpCDF	0.01
OCDF	0.0001

In evaluating dioxin analytical results, the mass of each of the congeners is calculated and multiplied by its TEF to determine the total equivalency quotient (TEQ) for a sample. It is the TEQ that is reported in most dioxin analyses, and this TEQ more accurately portrays the real toxicity of the dioxin within a sample, rather than if merely the total mass of all dioxin congeners is reported. As can be seen in the table, some of the congeners have a very small TEF, so even a relatively large quantity of one of them in a sample would not add much to the reported TEQ.

organisms in the manuals. In the bioaccumulation tests, labs were not just counting the number of surviving organisms (with or without the help of a scope in the case of the very small species). For these tests, organisms needed to be meaty enough for a lab to be able to do a chemical analysis on them. At least fifty grams of biomass are necessary to do the full array of chemical analysis, about half that to do dioxin alone. For crustaceans, mollusks, or fish, this would not be a problem. There were a number of species available that were easily large enough for chemical analysis. But one of the tests had to be done with zooplankton or phytoplankton, and this was another matter entirely.

Phytoplankton (plantlike plankton) were ruled out because the test aquaria would literally have to be soup to get enough biomass from these for analysis. Zooplankton can be larval forms of larger marine organisms or tiny, mostly free-drifting animals that feed on phytoplankton or other bits of organic detritus in the ocean. An organism had to be found that was a real zooplankter, would survive long enough in the test aquaria, and could generate enough biomass to do the chemical analysis. After careful consideration of a number of species, it was decided that the humble brine shrimp, common fish food for amateur aquarium keepers everywhere, should do the trick. To get enough biomass from the tests, the aquaria were in fact quite soupy, but the whole thing worked well enough to get the data the judge said were required.

Meanwhile, the sand capping was going, and going. The requirement for a one-meter-thick cap on all dumped material was proving to be a nightmare, costwise, to the PANYNJ. In his July ruling, Judge Debevoise had indicated that he was unsure of the suitability of the dredged material for ocean dumping but that the capping required as part of the disposal conditions alleviated his concerns. The PANYNJ was paying dearly for that peace of mind. The normal rule of thumb used by the Corps for capping operations was a two- or

FIG 4. (*Left*) The EPA Ocean Survey Vessel *Peter W. Anderson.* This photo shows the narrow hull design of the *Anderson* and its crowded deck when ready to deploy guard buoys and other physical oceanographic equipment. (Photo provided by Paul M. Dragos)

FIG 5. (*Below*) Guard buoys being loaded aboard. Railroad wheels can be seen on deck. These were used as disposable anchors for the buoys. In the foreground, the yellow sphere is one of the S4 current meters that were deployed at several depths during the winter surveys. (Photo provided by Paul M. Dragos)

FIG 6. (*Above*) Deploying anchors. The railroad wheel anchors were let out first; then the instrument arrays were attached to the cable. The instrument pods seen here in the foreground are much smaller than the large quadrapods described in chapter 3. (Photo provided by Paul M. Dragos)

FIG 7. (*Right*) Nansen bottles. Among the oldest (and still widely used) of all oceanographic instruments are these bottles for collection of subsurface water samples. They are generally metal or plastic tubes with plug valves at each end. Generally, a number of bottles are attached in series, with ends open, at predetermined intervals along the wire and lowered to the desired depth. They are closed in succession by the tripping action of a messenger (a small metal weight) slid down the wire. (Photo provided by Paul M. Dragos)

FIG 8. Video sled. Cable-towed real-time video sleds represent one of the newer technologies in oceanographic instrumentation. This one is towed close to the bottom—others, such as one used during the winter surveys to view the bottom at 120-foot depths for a new disposal site more than 20 miles from shore are remotely operated by cable. (Photo provided by Paul M. Dragos)

three-to-one ratio of cap material to dumped material, and that was what the Corps had indicated to the PANYNJ. There were two factors in this project that basically threw that rule of thumb out the window. One was the condition requiring the one-meter-cap, religiously, over the entire footprint of the disposal. And the precise methods being used to monitor this operation ensured that even the farthest few centimeters would be detected and included.

The other factor was actually a twofold one: (1) the material was to be disposed of in a series of flat mounds to a height of no more than sixty-five feet below the surface, which meant a larger than usual surface to cover; and (2) the material was very muddy and tended to spread out considerably. The sixty-five-foot depth requirement came out of the previously discussed poststorm monitoring project. Put together, these factors resulted in a much more spread-out footprint than the Corps' Waterways Experiment Station had initially estimated. The result was that large amounts of sand needed to be continuously dredged from another offshore site and barged over to the disposal mounds. Operational

FIG 9. Port dredging. Historically, an average of 4 million cubic yards of sediments have been dredged from the Port annually, ranging up to 7 million cubic yards in recent years from deepening projects. *Above,* the dredge "Chicago" working in Kill Van Kull. *Below,* a dredge working in Newark Bay, outside the Port Newark/Elizabeth Marine Terminal. (Photos courtesy of the U.S. Army Corps of Engineers)

FIG 10. Split hull barge. This type of vessel is typically used for most ocean dredge disposal operations. (Photo courtesy of the U.S. Army Corps of Engineers)

costs were also increased by the method required to lay the cap material. Split hull barges were required to steam slowly under strict navigational controls and "sprinkle" the sand over a prescribed area to ensure a good, even coverage (figure 10). These very controlled conditions took more time than usual disposal operations that include capping, and, as might be surmised, operational time costs operational money.

All in all, this was turning out to be the most expensive dredging disposal project in history, and the PANYNJ was not happy about it. Halfway into the operation over 2 million cubic yards of sand had been used to cap a 500,000-cubic-yard project, and it was only about half done! Why were the Corps' earlier estimates so much lower than what the reality was turning out to be? There was general sympathy and commiseration within the Corps and EPA for the Port Authority, but after all they were big boys who had their own engineers and had been doing dredging operations for a long time. There was enough general fatigue and frustration to go around among all concerned by this time that only a limited amount of sympathy could be generated for the Port Authority.

8

LET'S TEST AGAIN

LIKE WE DID LAST SUMMER

The Port Newark/Elizabeth project was not the only one around, however. During this time, EPA and the Corps had been locked in battle over the issue of the backlogged projects that had been held up for processing by the Corps. The original forty or so projects for which the amphipod or dioxin tests had not been conducted were still on hold in July 1993. A NOAA National Status and Trends study of the sediments in the Port had recently been completed, and it included amphipod testing results. The data from this study would be used to help determine which additional areas in the Port might be toxic to amphipods. For dredging projects in these areas, the amphipod test would always be required. Dioxin analytical results were also becoming available for the Port area from other sources, including studies conducted at Rutgers University. EPA used this information to identify the projects that were in areas of the Port complex that warranted these tests. It provided a sound basis for requiring the tests, targeting them only for projects in known problem areas. A letter was sent to the Corps in early July that identified the projects and the tests required, as well as the basis for the requirements for each project.

Test results were starting to come back for the projects that the Corps had been holding pending clarification of the testing, and a lot of them were failing the amphipod test. Recall that a failure of this toxicity test meant no ocean disposal, with or without capping. The Corps requested that EPA review the list of projects it had submitted for amphipod testing or reevaluate the passing criteria—anything that could provide a little slack on the failure rate. But EPA Region 2 could not retreat from its support of the test and from requiring it for the known areas of concern. All the other coastal regions were by now using amphipods for ocean disposal testing. The environmental groups were keeping a close watch on the regional testing program. Finally, these results should not have been a major surprise in any case.

In 1991, EPA's Narragansett lab had conducted some pretesting of the new proposed procedures, including amphipods, using Port sediments. The indications were that the revised procedures would result in significantly more failures and more Category 2 determinations than were then the case. Up to that time approximately 95 percent of projects met Category 1 requirements, and about 97 percent qualified for at least Category 2. There were only two locations with "sediments from hell" (agency staff colloquialism for the most contaminated sediments) in those days that were Category 3 in the Port area proper: Gowanus Canal in Brooklyn and Newtown Creek in Queens. These areas still have not been dredged in at least thirty years, partly because of the limited commercial use of the waterways but also because disposal of these sediments would be problematic. Both of these areas are currently being investigated for waterfront rejuvenation, which would almost certainly require dredging (and decontamination).

Back to the grim realities of 1991: the Narragansett lab findings indicated that the Category 1 and 2 percentages would probably be driven down to the 50s and 60s range if the new tests were implemented. Using these findings, EPA regional staff (author) had written a memo for management in November 1991 that forecast in some detail the probable increases in project failures that would result from implementing the testing revisions. These findings were also no secret to the Corps. So both EPA and the Corps knew, or should have known, that what was happening had been a foregone conclusion. The inescapable fact was that much of the Port contained sediments that would be considered too contaminated for ocean disposal by most scientists in the field. The fact that the agencies and the PANYNJ had not been prepared and had not developed alternatives to ocean disposal was no reason to blame the tests. There was a lot at stake, though, and the Corps and the dredging community would not give up so easily.

Other things that were beginning to happen around the end of the summer of 1993 had to do with the dredging forum. As was noted earlier, workgroups of government and nongovernment experts had been appointed to work on specific issues of the Port dredging problems, which included identifying non-ocean alternatives, managing the ocean site for eventual closure and looking for a new site, investigating decontamination technologies, and developing updated criteria for testing. An interesting aspect of these workgroups is that government scientists worked hand in hand with representatives of the environmental groups that were suing them. In the Criteria Workgroup, for example, issues regarding evaluation criteria for dredged material were being critically discussed while it was this issue that the agencies were being sued over. While the environmental groups were filing against the government for discovery, and the government was trying to document for the court its scientific basis for the testing, both parties were exploring some aspects of the technical details together openly in a workgroup! As some of the other ocean dumping coordinators from around the country were heard to say, only in New York! Somehow, though, despite these very apparent conflicts and the wide divergence in philosophy and personalities in this workgroup, progress was made on many of the issues that were tackled. This was even more surprising as word came in from some of the other workgroups that were not involved in the legal issues that they were self-destructing from disagreements and infighting.

In early October, the Corps called a meeting on the amphipod "problem," bringing in people from its headquarters and Waterways Experiment Station scientists. EPA brought in researchers from the Narragansett lab. The old questions were raised about whether the failures of bioassay tests were really due to contamination or whether there could be other factors causing failure such as ammonia, particle size differences, or other "artifacts" not related to the actual contaminants in the sediments. Various technical information was discussed, and in the end Narragansett scientists indicated they would provide appropriate water exchange rates for the tests. They now agreed with Corps scientists that allowing a minimal water exchange would help release ammonia while not compromising the evaluation of other contaminants in the sediments. Even though the standard test was "static" (no water exchange), a flow-through water exchange test would be allowed for this region.

Near the end of October 1993, Judge Debevoise granted the federal agencies an extension to complete their responses on the issues that he had identified. Affidavits were being completed to be included in the response package. The EPA affidavit was submitted by Mario Del Vicario, chief of the Region 2 Marine & Wetlands Protection Branch. It went into scientific detail about why the suspended-phase bioaccumulation test was not warranted, why an additional liquid-phase bioassay was not necessary, why one indicator species was sufficient for the dioxin test, and why the Green Book should be the determinant in cases of apparent conflict with the regulations.

Meanwhile, information was coming back from EPA headquarters that its dioxin reassessment human health document was almost completed. The results of reassessing all the information available, including new epidemiological data, indicated that dioxin was a complete carcinogen in animals, but the evidence could demonstrate no more than a "promoter" effect in humans. In other words, dioxin could promote, or help set up, the conditions in humans for cancer to begin, but there was not sufficient evidence to indicate that dioxin could actually cause cancers. However, the new results also indicated that dioxin's potential for causing immunosuppressive, developmental, and reproductive effects through its ability to disrupt endocrine systems were greater than had previously been believed. These findings cranked up the level of concern about dioxin another notch or two.

About this time, the EPA regional administrator's office pronounced that ocean disposal at any regional site, either the existing Mud Dump or a new ocean disposal site that might be designated in the region, was a method suitable only for Category 1, the "cleanest" material. The questions being raised regarding the effectiveness of capping contaminated material in the coastal ocean might have eroded some of the confidence the regional administrator, and EPA in general placed in the process. The Corps was not happy about this turn at EPA, since it would leave the Corps in an even worse situation than it had anticipated for needing a major disposal alternative. Even if borrow pits, some upland sites, and some decontamination technologies could be established in the next few years, these would probably not be enough to meet the Corps' or its applicants' long-held dredging needs. These alternatives could be used for the more contaminated materials (the latter two for the most contaminated), but there would still be large amounts of material that would not meet the Category 1 criteria. There had been some discussion at EPA about whether Category 2 material could be further subdivided into 2a and 2b, for example, with 2a perhaps being allowed for borrow pit disposal. The discussion did not go beyond this point, however.

The Corps strenuously objected to EPA about its decision. EPA felt that it was in the right and had the regulatory authority to manage or designate an ocean disposal site with any conditions it believed to be appropriate. Though this decision, like any other on dredged material disposal, would of course be subject to an elevation process initiated by the Corps for resolution at higher levels, this option was not pursued. In any case, as a policy issue, the EPA decision was not clear-cut, as it goes to the very heart of the ocean dumping criteria.

As was alluded to earlier, the ocean dumping criteria are not, as the term "criteria" connotes, numerical contaminant thresholds that can not be exceeded. There are in fact no usable numerical values in the Ocean Dumping Regulations, only narrative text of what kinds of tests to do and the qualifier of no "unacceptable adverse impacts" resulting from the disposal. EPA and the Corps are supposed to quantify this qualifier from test results, with EPA ostensibly having the last say about compliance with the "criteria." EPA regional management was basically saying that from now on, material that would normally have been considered acceptable for ocean disposal, albeit with conditions requiring capping, was no longer going to be allowed at all.

It was noted that the justification for the capping requirement that had historically been cited was that capping would provide an added level of protection from any uncertainty regarding a material's potential to cause bioaccumulation. The material still had to be "acceptable," however, or the regulations would not permit its disposal. If Category 2 material was now determined not to be suitable for ocean disposal, then it logically must be for one of two reasons. Either it was now determined not to be "acceptable" under the regulations, or a policy decision had been made that capping was not adequate to "cure" the potential for adverse effects. Neither of these was enunciated, though; the decision was simply made that Category 2 could no longer be disposed of in the ocean.

It was not clear to the Corps how a change from "acceptable" to not acceptable was justified, as there was no discussion given of additional information being available or further analysis having been made of the existing information. The other possible reason, that capping was no longer considered to be an effective mitigation technique, could have been made as a policy decision without a lot of technical analysis, given the results of the nor'easter along with other previous indications. But there was no assessment given as to why Category 2 was no longer determined to be acceptable, not only for the Mud Dump but for any other ocean site that would be designated in the region. It was not surprising, then, that the Corps was quite upset over this turn of events.

The regional dredging/ocean disposal program had now pretty much turned full circle, from a time when the Corps ruled basically without peer to one in which EPA was making unilateral decisions and essentially calling the shots. (The situation has apparently evened out a bit more lately, especially after the change in administrations following the national election in 2000.)

Although there were benchmark (matrix) numbers that the agencies agreed to use in the interim, it was the development of updated bioaccumulation thresholds to distinguish between Category 1 and 2 that was the main and most elusive effort of the Criteria Workgroup. This goal was (and apparently continues to be) elusive because establishing such thresholds relies on a risk assessment approach that requires a lot of input information that is often hard to come by, and it is a process that is infused with potentially subjective points of assumption. Scientific data relating effects to concentrations of contaminants in organism tissue are scarce for most contaminants, and so a stepwise series of assumptions must be made regarding toxicity data, exposures, feeding behaviors, food chains, and other factors that allow for subjective opinion at many points along the way.

A meeting was held by EPA Region 1 in Boston to try to come up with bioaccumulation thresholds for dredged material tests. Leading scientists in the field from academia, private research groups, and government agencies were assembled. After a full two days of sustained deliberation over the relevant data and methods available, in the end no consensus was reached on a threshold value for even one contaminant.

So, development of bioaccumulation threshold values would be difficult to attain even by a group of people who got along fairly well. There were members of the original Region 2 Criteria Workgroup's—bioaccumulation subgroup who could barely stay in the same room together. The subgroup had members representing EPA, the Corps, the states of New York and New Jersey resource agencies, environmental groups, and the dredging industry. Yet eventually this group began developing consensus on some general thresholds for the major contaminants. These were not precise one-number threshold values (the jargon is "bright line numbers"), but they were in a fairly well-defined range that could be narrowed by continued discussion and assessment of other information that was under review. A scientific peer review process had been planned for when the effort was completed. The group's proposed approach would most likely have acquired broad scientific approval. Unfortunately, the process was too slow for the fast pace of events that were unfolding.

EPA management decided that a simple, programmatic protocol was necessary to enable making decisions on the projects that were being tested right then. So, it sought to institute criteria that were based on a fraction (10 percent) of the existing FDA action levels (see chapter 2). The environmental groups protested against this as arbitrary and without merit. Their protesta-

The ideal scientific data to use in the development of bioaccumulation criteria would be from laboratory tests that gauge specific adverse effects (such as repro-ductive inhibition or larval malformations) on the specific organism in question and link those effects to measured levels of a contaminant in the organism's body. This type of data would be needed for all the contaminants that are typically tested for, about fifty or sixty, although some exist—and can be evaluated—in groups. Unfortunately, not a lot of this type of information exists, partly because the data have to be generated through experiments involving costly chemical analysis of significant numbers of replicate and other quality control samples.

There are laboratory data available that were used in deriving the EPA's and some states' water quality criteria some years before this. The experiments were done by exposing organisms to known concentrations of a contaminant mixed into the water used to expose the organisms. Virtually none of this information in-cludes data on the concentrations that were assimilated into the bodies of the or-ganisms. Also, the organisms were of a wide variety but not typically of the kinds used in the dredged material tests. There are also some data available on effects related to organism tissue concentrations, and these are preferable to any others. This kind of data would be used for estimating the potential for ecological effects, optimally for the potential effects on the most sensitive species in the target food chain.

The other component of a risk evaluation is human health effects, which can use exposure and food chain information similar to that used in the "ecological" component above. At the next step, after the potential contaminant concentra-tions of a human food source are estimated, any available human health effects data are coupled with additional factors to account for variables relating to the populations at risk and their consumption of seafood. For both ecological and human health assessments, a risk approach requires that values be assigned as appropriate safety factors. These would account for such things as differences in biological responses between different species and differences in responses among individuals in populations of target organisms.

Also, adjustment factors have to developed for actual time of exposure/uptake in the field compared with test durations (steady-state factor), for food chain (trophic) transfer of contaminants, and sometimes for whole organism versus filet concentrations to use in consumption estimates (humans do not usually eat the whole organism). To do this, scientific literature and any other relevant informa-tion is identified and evaluated to quantify such factors to the best degree pos-sible. With respect to humans, things get more complicated when trying to estimate ingestion rates of particular kinds of fish from specific areas and by specific human populations or subpopulations, such as pregnant women. All this is put into place and back-calculated to arrive at a threshold concentration in the body of the benthic test organisms that are used in the dredged material tests.

tions were based firstly on the fact that EPA itself had argued against the FDA value for dioxin as being not appropriate. The other FDA levels were considered, in the Green Book, to be appropriate only as upper-level ceilings for regulatory decision-making. Further, the 10 percent threshold was admittedly arbitrary and without any real basis for this purpose. The new criteria were summarily withdrawn, and the old "matrix" values mentioned earlier were put back in place.

These matrix values had been developed jointly by EPA and the Corps in the early 1980s and had been used ever since for bioaccumulation evaluations. The values reflected the "grand mean" of tissue concentrations of four major contaminants (PCB, DDT, mercury, and cadmium) in various marine species in the Mud Dump Site surroundings. As was noted, a few of the matrix values were calculated from measured ocean water concentrations of these contaminants in the area, owing to lack of biological field data. The matrix had individual values for each contaminant for each of the two major test species, worms and clams. Though the values were developed without any consideration of effects levels, they were fairly conservative and protective since they represented the average concentrations for organisms in a wide swath of the ocean, including areas far from the disposal sites within the Bight Apex. (The Apex is the inner part of the V of the New York Bight, located at the entrance to the Port.)

The principle that was applied in using the matrix values for bioaccumulation test evaluations was termed "no further degradation." This meant that if bioaccumulation test results were no higher than the matrix values, the tested material would not degrade the disposal site or the environment any more than was already the case. This approach had been planned to be only interim, however, since it was known even then that there were other factors that should be brought into play, including consideration of effects and risk, and that therefore the matrix numbers would be updated in due time.

The Criteria Workgroup–bioaccumulation subgroup continued along on its bumpy road for some time, developing several preliminary criteria and methodologies that could be used in subsequent work. The NYSDEC representative, John Zambrano, chief of the Standards and Special Studies Section in the Division of Water, wrote the risk assessment report for the group that included a derivation of proposed human health criteria and wildlife criteria. For these derivations, the group had developed applicable human and wildlife consumption rates, trophic transfer factors, whole body/filet ratios, lipid adjustments, and a representative food chain at the ocean disposal site. Zambrano also developed a conceptual model and computation steps for deriving the criteria, based on the principles of water quality criteria derivations. Much of this work was used directly in subsequent efforts (see below). The work was not ultimately completed, however, and the bioaccumulation subgroup eventually disbanded. As the EPA representative and coordinator for the group, the author presented the work that had been accomplished at a national bioaccumulation conference in Washington, D.C., in 1996 (Lechich 1998). There is an ongoing criteria development effort managed by EPA under a different name, the Remediation Material Workgroup (RMW), at the time of this writing (early 2005).

The Corps had decided that it needed to demonstrate its displeasure with EPA's decision allowing only Category 1 material into the ocean. There was an ongoing program of ocean surveys and studies with the goal of identifying alternate ocean disposal sites, funded jointly by both agencies. The Corps had committed $150,000 from its annual budget allocation for its share of this work. When EPA banned Category 2 materials from the ocean, the Corps decided that further oceanographic work under these circumstances was no longer in its interest, and it pulled its funding. EPA staff had to reshuffle the planned oceanographic work to compensate for this, but they were not surprised by the Corps's reaction. Besides, there were other things that the two agencies were still jointly involved in, and after all they were still codefendants in a major lawsuit with potential national repercussions.

When a lawsuit of national import occurs in an EPA or Corps regional office, it is not usually long before headquarters gets involved. Besides the fact that both agencies' guidelines essentially require this involvement, such lawsuits also provide headquarters the opportunity to demonstrate how helpful it can be. The lawsuit initiated by environmental groups had national import for two reasons. First, it was known that at least one of the groups, Clean Ocean Action, had gained an audience with the New Jersey governor's office, and rumor had it that COA was gaining access to the White House Council of Environmental Quality and Vice President Gore's office. Second, the lawsuit specifically challenged the legality of the national Green Book, which was the foundation for EPA's and the Corps' national dredging program. Losing this lawsuit could have implications on what both agencies' headquarters had been doing in this regard for the last ten or more years. So, Region 2 EPA and the New York District Corps of Engineers were very much at the center of the national dredging program's attention, with the other coastal regions around the country looking interestedly on. In reality, this focus had been the case even before the lawsuit, due to the dioxin and amphipod issues in the New York region.

In early November of 1993, one of the forum workgroups, the New Site Designation Workgroup, suggested that Category 1 material should be used to remediate contaminated areas around the Mud Dump Site. One of the more recent surveys had discovered that there were more contaminated sediment areas in and around the site than had previously been thought. Sediments were brought back for labs to do the updated bioassay tests, and the results were quite worse than expected: sediments in a number of areas in and around the Mud Dump Site were equivalent to Category 2 or even Category 3 sediments. There was a lot of nail-biting about what to do. One option that was provided in the Ocean Dumping Regulations, as noted before, was closing the site to any further disposal and commencing remediation activities. Of course, closing the site was not palatable to either agency, but someone in the New Site Designation Workgroup hit on an idea that could kill two birds with one stone. Instead of closing the site and designating a new site (the workgroup's task), why not cap these more highly contaminated areas with Category 1 sediment, thereby remediating the site and saving the trouble of designating a new one? The idea quickly caught on and was endorsed by the decision-makers in both agencies, and planning was started to carry it out.

Only two weeks after this decision, some shocking news came from the New York District. It was discovered that, somehow, a mound had accumulated at the Mud Dump Site that projected upward to only about twenty feet below the surface! (This in an area of the coastal ocean normally about eighty to ninety feet deep.) The mound was made of material from an ongoing Corps federal deepening project in the Kill Van Kull and was formed apparently because of the requirements for more accurate dumping. Disposal barges had kept dumping at the exact spot they were told to by the Corps. Apparently the operations were not being kept track of well enough (including by EPA, which by now had staff assigned to coordinate with the Corps on site management) for anyone to realize that a huge mound might be forming. The mound was judged to be an immediate navigational hazard, and a dredge was soon dispatched to dig it out and redistribute the material. EPA fired off a letter to the Corps that talked about the need to upgrade site management procedures.

At the end of December 1993, Judge Debevoise heard arguments on the federal defendants' motion seeking relief from a broad-ranging discovery request by the plaintiffs. The judge reserved decision on that motion (later granting discovery in specific areas). In the course of the hearing, the judge responded to the government's queries about his July court opinion and the Corps' ability to issue other permits. He said that he had not intended to stop any further permits but intended only that they be issued in compliance with the Ocean Dumping Regulations. Since it was in fact the interpretation of these that was the center of controversy in the lawsuit, the agencies remained somewhat mystified as to the judge's reasoning or intent by this proclamation. Nevertheless, the Corps used it to justify a decision to once again issue permits, and it authorized seven projects that were believed to have completed adequate testing.

As for the other thirty-eight or so backlogged projects, EPA and the Corps seemed to have more differences within their respective internal offices than between the two agencies. Both EPA and Corps counsel, taking a conservative view that sought to protect their agency against legal sanction to the greatest extent, suggested requiring the full suite of tests that had been done for the Port Newark/Elizabeth project. The Corps and EPA program offices at the region, district, and headquarters levels believed that the logic of science should prevail. They believed that EPA should require only the tests that were scientifically necessary to characterize the material, and that they should keep making their case to the court. A policy developed that was sort of "don't ask, don't tell." At all the subsequent meetings with applicants, the applicants were told that EPA would not require the additional testing in order to make a determination on their project. At the same time, both the Corps and EPA cautioned them that it was a "business decision" on the applicants' part whether to do the additional testing, since not doing so could place their projects in legal jeopardy.

In a continuing spirit of cooperation, EPA and the Corps signed a new memorandum of understanding on how to jointly manage the Mud Dump and to coordinate on project testing and evaluation. The discovery of "Mound Everest" (EPA and Corps staff had a different term for the mound that mocked a certain manager) helped to grease the Corps' skids on agreeing to the memo. It called for much more proactive participation by EPA in almost all areas of dredged material management. It included a more active role in site management responsibilities. EPA Region 2 also now had what it had long been seeking: early coordination on Corps federal navigation channel and regulatory project applications in order to work out appropriate sampling/testing schemes jointly. This was important because prior to the agreement EPA was often able to comment on projects only after sampling and testing had already begun, which was obviously a weak position from which to negotiate. As part of this arrangement, detailed testing requirements regarding ammonia in the amphipod tests, developed through EPA headquarters, were also adopted into the testing regime. (It may be surprising to the reader that the ability of EPA to jointly review projects for sampling and testing did not happen until 1994, a full two years after establishment of the revised regional testing manual. Credit for this should be given to the prodigious stonewalling skills of the New York District Corps and to the curious proclivity of EPA Region 2 for appeasement on this issue.)

Another problem that was developing with the amphipod test was now being raised by a regional laboratory that had been testing Port dredged material. It was claiming to be having problems finding the right-size organisms in the local waters during the winter. It also said that the organisms grew beyond the required size for testing during the time required for test conditions to meet the new ammonia requirements. Lastly, it was having trouble meeting the reference requirements, which specified the minimum amphipod survival level that had to be achieved using "clean" reference sediments. A major technology transfer effort was launched between this lab and EPA laboratories in Narragansett and Edison, New Jersey, to address the situation.

The U.S. Navy had been patient thus far, waiting to see what decisions would be made on testing because it had a major dredging project of its own. Naval Weapons Station Earle, near Sandy Hook, has a huge docking facility that is the largest structure in Raritan Bay. (Raritan Bay is in the lower harbor, near the entrance to the Atlantic Ocean.) When the navy representatives came in to meet, they were informed that EPA headquarters was on the verge of issuing a draft regulation that would cure the remaining testing ambiguities. A slew of meetings with other applicants followed, primarily with the major petroleum companies in the Port, including Exxon, Hess, Chevron, Northville-Linden, Citgo, and Mobil. Other dredging projects that were on deck included

Particular size ranges of amphipods were required for toxicity tests because males of an advanced age (and size) tend to die abruptly due to sex-linked traits and therefore would confound the test results. Therefore, a size range typical of juveniles of either sex was required when field-collecting the organisms. The new ammonia protocols required that test sediments collected from a project site had to be placed in the test aquaria with changes in seawater and measured for ammonia until the ammonia concentrations fell to acceptable levels. As it was turning out for some projects, this time period ranged up to twelve or fourteen days (although it was usually much less). The lab that was having difficulties performing the tests claimed that it had to collect organisms as soon as project sediments were received, because the lab staff never knew how long this purging period would be. (The amphipods had to be acclimated to the lab conditions for a couple of days before they could be used.) By the time these long purging periods were over, it was claimed the amphipods had grown to beyond the required size! They also had problems finding the right-size ones in winter, claiming juveniles were not present during cold water periods, due to their breeding cycle. The EPA scientists ruminated over these problems and proclaimed: (1) there were juveniles overwintering in Narragansett Bay, which was colder than New York Harbor; (2) the timing problem could be worked out with a little premeasuring of ammonia in project sediment samples; and (3) EPA Edison laboratory had had no problem in consistently achieving good reference survival and would help the contract lab to do so.

Since much of the contract laboratory's problems were due to field-collecting the test organisms, EPA could have required it to obtain organisms from one of the several commercial sources that were becoming available. The tiny amphipods were costlier than jumbo shrimp (their taxonomic big cousins), and it was cheaper for testing labs to collect their own from nearby bays. Although EPA was being buffeted by the enviro groups from one side, it also had to demonstrate to the regulated public that it was being as fair as possible by allowing as much choice as possible regarding sources of test organisms.

A new twist was being raised by the Corps regarding the ammonia problem. What was good for amphipods must also be good for mysids, the other solid-phase toxicity test organism. Mysids are also crustaceans that have long been used for water and other testing, but they, unlike some amphipod species, are filter feeders. Recall that the Ocean Dumping Regulations require representation in tests by burrowers, deposit feeders, and filter feeders. Burrowers, as might be imagined, feed on organic items in or on the sediment that they ingest as they burrow through sediments. Deposit feeders for the most part eat the organic material that has been deposited on the surface of the seafloor. Amphipods (of some species) can do both of the above. Filter feeders filter or trap material that is floating in the water column. Mysids can be deposit feeders or filter-feed while resting on the bottom or actively swimming, and many filter feeders are free-floaters, such as jellyfish and other similar organisms. Therefore, since mysids are both deposit and filter feeders, including them would satisfy the three feeding modes required in the regulations, and so they had been assigned as the other solid-phase toxicity test organism. Upon receiving a formal request from the Corps, Region 2 went back to EPA headquarters and worked out another guidance letter on appropriate ammonia protocols for mysids. The politics of the situation in the Port were by now such that any revised procedures in Region 2/New York District were being coordinated through headquarters. The other ocean dumping coordinators must have wondered, why is New York having all these problems?

bulk shippers, dry dock companies, and of course the PANYNJ and Corps federal projects. The private projects were all being advised on the don't ask–don't tell policy.

By the beginning of 1994, only 75 percent of the disposal mounds from the Port Newark/Elizabeth project had been capped with one meter of sand. The project was finally completed, capping and all, by February. It was estimated that the disposal cost for this project worked out to about $35 a cubic yard, as opposed to the historical average of about $4 per cubic yard for disposal at the Mud Dump Site.

The results of the additional (court-ordered) tests done with the other test organisms for Port Newark/Elizabeth indicated no increased level of concern compared to the original tests. The project was still deemed to be a Category 2, acceptable for ocean disposal with capping. In June 1994, Judge Debevoise issued an opinion that it would be unlikely for the plaintiffs to prevail on the merits of any of their claims. He ruled that the defendants had met the regulatory requirements by conducting the additional tests he had "requested" and that the agencies had wide discretion in determining which tests should be conducted and the manner of conducting the tests. He formally denied the plaintiffs' request for a preliminary injunction. In August, the plaintiffs appealed his decision to the U.S. Court of Appeals for the Third Circuit.

To add some geographical perspective for understanding the scope of the testing revisions, it should be noted that the Port of New York and New Jersey was not the only major dredging area under EPA Region 2's jurisdiction.

Organisms that are used in solid-phase tests, both toxicity and bioaccumulation tests feed mostly on organic matter in or on sediments. A little more explanation on this is called for, since organic matter is such an important characteristic of sediments. The origin of most coastal sediments along northeastern American shores is from erosion of coastal land formations. Rocky headlands are eroded and form sand particles that are made up mostly of quartz and other hard, dense rock material. The coastal ocean is a fairly energetic environment, with large storm-induced waves having the ability to create conditions on the bottom in which finer and lighter particles will be resuspended and carried off. Sand particles, because of their density and weight, are harder to resuspend and transport. That is why most areas on the inner coastal shelf have sandy, larger-grain-size sediments.

There are some areas that have finer sediments, but, except for deeper areas such as the Hudson Shelf Valley, most of these are thought to be transient features that result from particular current and wave patterns associated with a single storm event. Or, during a snapshot in time, they might be relics of such forces over a longer period. The coarser sandy sediments usually have only small amounts of organic matter, because organic matter is less dense than the sands and tends to wash away rather easily. There are other physical-chemical factors associated with binding of organics to sediment particles that also play a role.

In estuaries and bays, the sediments can be much finer and have much higher amounts of organic matter associated with them. Their origins are mostly from rivers that flow into the estuary, and these carry fine silt material from soil erosion of upland areas. Included in this suspended load is organic matter that can have many sources, including the soils, which can be made up largely of previously living plant material that has decomposed. This decomposed material, along with that of animal origin, is termed organic since it was once living matter; and since living matter on this planet is based on the element carbon, it is measured in terms of organic carbon content.

The fine suspended particles carried by rivers and streams have a greater ability to bind with organic material than do sand-size particles, because of molecular attraction forces between the fine particles and organic matter. Benthic organisms, as mentioned before, feed on the organic matter in sediments and therefore tend to be much more numerous in silty sediments than in sandy ones. One problem that had to be dealt with in developing test protocols with amphipods was that two of the species more suited for testing were also the ones that preferred more organics-rich, silty sediments. When using these organisms for silty project sediment tests to be compared with tests using the same types of organisms exposed to sandy reference sediments, differences in responses due to these preferences had to be taken into consideration. The early developmental tests helped to quantify these variances as well as identify the best species to use according to the project and reference sediments being used.

Molecular forces similar to those responsible for the preferential binding of organic material to fine sediment particles also apply to contaminants. Fine particles, as is the case with all small bodies in comparison to larger ones, have a larger surface-to-volume ratio than sand size-particles. What this means is that, besides the molecular attractive forces at work for both organics and contaminants, the larger surface area per volume of the finer sediments allows for much more bonding to occur. This is why fine silty sediments can have both more organic material, which is an attractant for benthic organisms, and higher contaminant concentrations, which can be picked up by the benthics (bioaccumulation) and transferred up the food chain, sometimes in increasingly higher concentrations at successive levels (biomagnification).

The region also includes a large portion of Lake Ontario, with several industrial harbors in its jurisdiction, and in some quirk of geographic partitioning Region 2 in New York also ended up with Puerto Rico and the Virgin Islands in its jurisdiction. The regional dredging program was addressing major port dredging issues in Lake Ontario. These were coming under the umbrella of the draft Inland Testing Manual, which was modeled closely on the Green Book, and so there were similar testing issues in the process of being worked out with the Corps Buffalo District and others.

The sampling and testing requirements for both of these areas of the country other than the Port were somewhat different because the specifics were coordinated with different Corps districts and because the peculiarities of each area called for some differences in approach. These differences had mostly to do with using different kinds of organisms, either fresh water ones in the Great Lakes or other, more indigenous ones in Puerto Rico. But there were other differences also. A major federal dredging project in San Juan Harbor was tested under revised methods, and failed toxicity results had placed portions of that project in jeopardy.

In San Juan, the dredging project testing was developed with the Corps Jacksonville District, and it included much more testing of individual samples as opposed to composites of a number of samples mixed together. One reason for this was that the quality assurance/quality control (QA/QC) requirements for chemical analysis of both sediments and bioaccumulation tissues were much less exhaustive (and costly) than was the case for projects in the Port of New York. Therefore, all the analyses and bioassays in San Juan were performed on individual samples, and this may have had a considerable effect on the decision-making for this project.

The quality control procedures for chemical analyses of samples in the Port had been established for the regional testing manual by the EPA Region 2 Edison, New Jersey, laboratory QA/QC unit. EPA headquarters had told Region 2 that its QA/QC procedures were the most extensive of those in all the regional testing manuals. It was not hard to read between the lines—they were perhaps too extensive. Since it was these exhaustive and costly QA procedures that drove the testing costs so high, they were no doubt influential in the need for EPA Region 2 to accept more compositing of samples within a dredging project. This was because the costs were such that it was difficult for EPA to argue against allowing cost reduction measures, such as compositing. The New York District Corps' argued that even if less compositing were required (more samples tested individually), this would provide little more information about the potential effects of a project, because the sediments would all get mixed at the ocean disposal site anyway.

The argument against the Corps' position had several parts. First, the Green Book required appropriate characterization of a proposed dredging area, and if inappropriate (too much) compositing of samples occured, it could mask sediment contaminant hot spots by diluting samples from these areas with samples from less contaminated areas. This concern was independent of disposal sites or methods from one important aspect; the act of dredging an area resuspends sediments that can potentially harm the surrounding marine environment. If the sediments in a hot spot were resuspended in an area where they might affect sensitive marine resources, the effects might be more serious than would otherwise have been predicted because the practice of compositing for testing might have diluted out the hot spot sediment. Secondly, areas were dredged and transported to the disposal site in individual barges with capacities of several thousand cubic yards. Over the course of a large project of several hundred thousand cubic yards, there could be any number of individual barges containing potentially much more contaminated sediments than the "averaged" conditions presumed from inappropriate sampling and testing. If an unfortunate fate befell one or more of these barges, or if the material was dumped somewhere that was not adequately identified and capped, the potential adverse effects, again, might be worse than would have been initially projected.

Several large sections of the federal project in San Juan that had failed toxicity tests were the subject of continuing discussions for a time with the Jacksonville District. It requested retesting for some of the areas, because of the fact that only one of the two test organisms had failed. Normally this is not agreed to, since the guidance requires the testing of more than one species because some contaminants or combinations of them will affect different organisms differently. The testing is designed to account for this fact—that although certain sediments may not adversely affect one species, there could be others that it would. So a failure of one of the two solid-phase toxicity tests usually means a refusal for ocean disposal. In this case there were other extenuating circumstances, so a limited retesting was allowed, but even after this there still remained significant areas of the project that had failed.

In a situation like this, the Corps will typically reevaluate the need for dredging and might also meet with the commercial and public interests served by the particular federal channel and reexamine their options. If there remains a crucial need for navigational access after this process, other disposal methods may have to be identified. In San Juan, it was determined that the sections that had failed were not crucial for the present time, and they were put into abeyance.

A question might be raised whether projects in the Port were (and still are?) treated less strictly than in the Caribbean, since there could be portions of some Port projects that would have failed under the same circumstances. This no doubt relates to why the results from the first Mud Dump Site biological/chemical monitoring effort in 1990 were greeted with some relief. Although all projects that had been permitted for disposal before then had undergone testing, the uncertainty surrounding the condition of the site then surely reflected to some degree concerns about the testing, and compositing of samples was one obvious source for such concern.

The answer to the above question, though, is possibly. There is no question the chemistry results generated for Port projects, with the intensive quality assurance requirements, were more likely to be exact than in the Caribbean projects, which sometimes, for example, did not achieve low enough detection limits. On the other hand, conducting individual sample bioassays, especially the toxicity bioassays, definitely produces a better characterization of the overall project conditions. Should the Port sampling protocol be modified to better reflect these considerations? The author believes the answer is yes, for toxicity tests and sediment chemistry analyses, while at the same time reevaluating the current QA/QC requirements for the latter, and for bioaccumulation tissue chemical analyses.

By the fall of 1994, EPA Region 2 executives had become alarmed by more of the early findings of the national EPA dioxin reassessment. The cause of concern was that there was very little gap between the average levels of dioxin found in human body tissues and the levels that were estimated to result in an increased risk for cancer and other health effects. The question that arose was, given that slim buffer, could any increased exposure to dioxin be allowed from a government-permitted program? The question was put to the head of the reassessment effort at headquarters. The answer came that any increase of dioxin concentrations in humans would probably come predominantly from normal dietary intake, such as from standard agricultural, livestock, and dairy products that are typically consumed by the average American. This would tend to overwhelm any comparatively minor environmental source, such as from an ocean food chain in a geographically limited area.

Such a generalized view may not hold, however, when considering a subpopulation of fish eaters who obtain a portion of their food intake directly from fish that may be contaminated by the disposal operations in question. Indeed, because of concerns from local fishermen and other members of the public, EPA decided to fund the National Marine Fisheries Service to collect game fish from an area around the Mud Dump Site and analyze their tissues for levels of contamination. So a few scientists and technicians from the NMFS Sandy Hook lab got to go out and catch fish the old-fashioned way, with hook and line, and collect specimens of fluke, blackfish, sea bass, and bluefish for chemical analysis. There were delays caused by quality assurance problems with the NMFS contract laboratory's dioxin analysis, which in turn caused considerable exasperation among the environmental and fishing groups because the results were eagerly awaited. The contract lab's quality assurance practices were resolved, but some of the analyses had to be redone. Finally the tests were completed and the results issued to the public. The findings were received with considerable relief by the agencies and the public: the levels of dioxin found were not such as to warrant concern (many samples were at non-detectable amounts) in the four species sampled from around the disposal site area.

In October 1994, EPA headquarters issued a final rule that clarified the Ocean Dumping Regulations. Bioaccumulation testing for the suspended phase of materials was stricken from the regulations. This lifted a large weight from the agencies and the dredgers. Unfortunately, it was not a panacea, since there remained other contested testing issues relating to the number of species required.

Then, a major waterfront facility recently purchased on Staten Island by the PANYNJ, Howland Hook Terminal in the Arthur Kill, failed the amphipod tests. This was a major disappointment because a lot of planning and funding had gone into the slated redevelopment of this facility as a major Port terminal. The development of this terminal would finally bring back more of the Port's business to the state of New York and help balance the overwhelming capacity New Jersey enjoyed, owing mainly to Port Newark/Elizabeth. Something had to be done. A lot of pressure was brought to bear on EPA, but the testing results were clear and spoke for themselves. An alternative had to be found. The PANYNJ felt that it could spare no expense to dredge and to find a disposal site for the dredged material. Given the level of commitment for this project, it was expected that all resources would be brought to bear to bring it to fruition, but no one expected the enterprise to turn out quite the way it did.

After several alternatives were explored, the final answer was that dredged material would be placed on barges and shipped to Texas, from where it would be loaded onto railcars for transport to a disposal site in Utah. This amazing journey brought to mind the errant Long Island garbage barge in the 1980s that tried to off-load at numerous southern ports, only to have to return to Long Island to finally dispose of its load. There is a Howland Hook disposal barge story also. On the trip down to Texas, one of the barges ran into heavy weather in the Atlantic and tried to put in at the port of Charleston. Sadly, it foundered on a sandbar and spilled some dredged material into Charleston Harbor. When word got out that there was dioxin in the material (in Category 2 amounts, not toxic waste concentrations), the good people of Charleston became very concerned. Their hue and cry was raised to high decibels, and the PANYNJ and EPA scrambled to provide information and an assessment of the potential hazards to the city of Charleston and the Coast Guard. A full cleanup of all lost material was ordered. For the relatively small amount of material that was actually lost from the barge, it was reported that twenty times as much sediment was dredged up and disposed of, at a cost of about $7 million. The cost of the disposal operation as a whole averaged about $113 per cubic yard, spectacularly breaking the record set by Port Newark/Elizabeth for the most expensive dredging disposal project, for its size, yet known.

The testing failure of the Howland Hook project did not sit well with the PANYNJ, and continuing rumblings of "mistrust" of the amphipod test were heard from it and others. PANYNJ officials called EPA management in the early summer of 1995 and alerted EPA to expect a detailed letter and study that called the test into question. The PANYNJ had contracted a regional testing laboratory to run some tests by varying some of the test conditions that were claimed to be problematic. The Corps was not, apparently, directly involved in this effort, but it was keeping a close eye on it. The amphipod test was creating a problem for some of its projects as well. Finally, the letter with attached study results and conclusions came in July 1995.

An EPA regional scientist (author) was responsible for addressing the PANYNJ's concerns and evaluating the technical basis for them. A point by point rebuttal of the lab study conclusions was drafted and discussed in a series of conference calls with national EPA experts in Washington, D.C., Narragansett, and Oregon. The rebuttal was approved by the national experts and then by EPA regional management. It was then sent to the District Corps and WES for their buy-in. There was an expectant waiting period, but it was not long before the Corps stepped up to the plate and agreed to the entire rebuttal, as written. It was jointly signed by EPA and the Corps and sent to the PANYNJ. Apparently, the Corps' prompt cosignature on the rebuttal, a class act on its part that could reasonably be construed as being against its own self-interest, put an end to the matter as far as the PANYNJ was concerned. There was no further mention by them or any other party of the reliability of the amphipod test.

In June 1995, the U.S. Court of Appeals for the Third Circuit overturned Judge Debevoise's previous ruling, which had supported the PANYNJ and the federal agencies. Moreover, he was found to have committed a "serious legal error" in his finding, the prior June, that the defendants had met the regulatory requirements by conducting the additional tests he requested and that the agencies had wide discretion in determining which tests should be conducted and the manner of conducting the tests. The agencies were struck by the appeals judges' ruling. The appeals court statement was somewhat bizarre in how abstract and difficult to decipher it was. There was nothing in the language that explained the decision. It did not address in any way EPA's contentions regarding the species/feeding mode argument. It cited the "plain meaning" of the regulations and the need to conduct "bioassays" on the suspended particulate and solid phase with species as discussed in the regulations. The implication was that three species had to be tested.

EPA headquarters and the Region 2 program still felt that the number of species necessary to cover the three feeding modes discussed in the regulations was the scientifically appropriate and sufficient number (two). EPA counsel believed that there would be continuing liability, however, if the "plain meaning" of the regulations was not followed. The Corps indicated it could support using one more species for dioxin testing (a pure filter feeder) to partly address the three species issue of the "plain meaning." Several private projects, including the Exxon terminal on the Arthur Kill, had already done additional testing with a second species for dioxin and were threatening to sue if they had to go through the whole procedure once more. EPA counsel thought that a good, clear argument (presumably beyond that already submitted) was necessary for why two species could cover three feeding modes. EPA headquarters was requested to weigh in on the issue. Should Region 2 go with the judges' ruling (three species) or go with the Green Book (two)? Headquarters stuck with the Green Book but stipulated that applicants should continue to be made aware of the potential liability. It added that a pending additional regulatory change that would fix the situation was imminent.

The meetings with applicants that followed were all about trying to clarify the muddy legal environment that the dredging community found itself in. There was much protesting by the companies about the situation and the increasingly higher costs of testing. Some projects were costing a couple of hundred thousand dollars just to do all the tests.

The companies wanted to know more about any disposal alternatives that might be available, since these would not require the high cost of testing for ocean disposal. The Corps could cite only what everybody pretty much knew—upland disposal, decontamination, borrow pits; all these were being planned, but nothing was going to be available before at least a couple of years. Many of the applicants that had not already done so were told that they would have to test at least one additional species, a clam, for dioxin. The Northville-Linden oil terminal had gotten to a point where it was almost issued a permit but then was told it had to do the additional testing.

It was in November 1995 that news was received of the Howland Hook barge, mentioned earlier, that had gone astray. As a historical footnote, it was at this same point in November (from the fourteenth to the seventeenth, to be repeated from December 18 through January 5) that the federal government shut down as the Republican-controlled House lost its budget battle with President Clinton. Most federal employees were furloughed (with pay

EPA was saying that a reasonable interpretation of the regulations was that fewer than three species could be used to cover the three specific feeding modes. It was reasonable to assume that one species that normally took in food by more than one mechanism, say by filter-feeding and also by deposit-feeding, should be a fair representative of the potential accumulation of contaminants from both of those modes. Given the legal aspects of all this, however, hypothetical "devil's advocate" arguments were tossed around by the scientists and lawyers. A contradictory argument could be made that, though highly unlikely, could still be valid. Organisms that can typically feed in only one of the three modes could potentially accumulate more contaminant than ones that can go to an alternative feeding mode. A strict burrower, for example, could be more exposed to contaminants in a test sediment than an organism that can switch to deposit-feeding to avoid the potentially more contaminated deeper sediment. This could occur because of the surface sediment becoming relatively depleted in contaminant content through contact with the oxygenated overlying water, and it could ostensibly result in a less than conservative assessment of the bioaccumulation potential of the material. Such a scenario is possible but nevertheless somewhat of a reach. It is an example of how almost any aspect of bioassay testing can be questioned ad nauseum, but reasonable decisions have to be made using best scientific judgment and common sense.

for the most part) during these time periods, and most appreciated the paid time off. Nevertheless, with the majority of federal employees necessarily being deemed "nonessential" in order to be furloughed, many possibly felt the time off was somewhat tainted by a poor choice of terms used to describe them.

On December 11, 1995, on the heels of his castigation by the appeals court, Judge Dickenson R. Debevoise issued his final decision on the Clean Ocean Action lawsuit. The judge granted COA's motion for summary judgment, the result of which was to find the permit issued by the New York District Corps to be void. He found that the permit had been issued illegally. In the course of oral argument, solid-phase testing was discussed, but the court did not finally dispose of the issue. The atmosphere of considerable uncertainty within the regulated dredging community was thereby continued and heightened. EPA continued to maintain the verity of the testing requirements of the Green Book, which was consistent with testing practice nationwide. Headquarters now took the position that this was a local decision that applied only to Region 2. It felt that if EPA were challenged again, presumably it would have the opportunity to argue the scientific arguments again and this time perhaps to win.

EPA Region 2 counsel and program staff continued trying to work out a suitable strategy for testing in the face of the court decision. One option was to localize it even more than EPA headquarters did—two species would be supportable generally, but in New Jersey (the court's jurisdiction) three species would have to be tested. Headquarters noted that regulatory changes to fix all this were coming.

Two other events that brought the year to an end were reflective of the dredging/shipping situation in the Port of New York. One was the refutation by a judge of a preliminary injunction, very similar to Clean Ocean Action's request (and whose plaintiffs were in consultation with COA), to stop dredging at the Seawolf nuclear submarine station in Groton, Connecticut. The news of the Port Newark/Elizabeth lawsuit issues had spread, but although the Seawolf lawsuit shared a similar beginning it would not be decided in favor of the plaintiffs. The judge there ruled in favor of the government, partly on the technicality that the claimants did not file within the time constraints.

Then, in late December, the National Park Service called EPA with a last-ditch request for an important dredging project that was being stalled. The proposed dredged material from around Liberty Island, home of the Statue of Liberty, had failed the new tests for ocean dumping, and the question was, could EPA possibly help identify any alternative disposal option? The channel leading to the island was becoming too shallow for the cruise boats that brought the public and for service vessels. Access to this symbol of our nation's freedom was being threatened by the Port of New York dredging crisis! There was no immediate relief available, but a resting place for the dredged material was eventually found in a New Jersey land site.

In early February 1996, the *New York Times* interviewed managers at EPA Region 2 for one of a series of articles on the dredging crisis. (The series began with a March 18, 1996, special report by Andrew Revkin titled "Shallow Waters: A Special Report; Curbs on Silt Disposal Threaten Port of New York as Ships Grow Larger." A search of *New York Times* archives going back to early 1996 reveals at least twenty-two articles on the dredging crisis from early 1996 to early 1997, with a few additional articles listed that have some connection to the Port dredging crisis.) The *Times* asked how the crisis had come about, what was being done, etcetera. EPA Region 2's responses included hard-copy overviews of dioxin issues and other testing matters. The Port dredging crisis had finally made it onto the pages of the Old Grey Lady, although many articles on the crisis had appeared much earlier in local New Jersey newspapers.

Finally, in late February 1996, EPA administrator Carol Browner signed a proposed regulatory "clarification" that addressed many of the outstanding testing issues. The final regulation did not become effective until September, and it was much reduced from the proposed rule because of strong opposition from the environmental community. The only final clarification was nevertheless an important one, that only two species were necessary for the solid-phase tests. The new regulation still left a lot of potentially interpretable areas in testing, but it did put to rest probably the most important issue from the federal agencies' and dredgers' points of view. The testing cost savings from this would be considerable, since it is the solid phase that requires bioaccumulation testing, with its extensive analytical costs. More importantly, from the agencies' point of view, it brought the regulations into conformance with the Green Book and prevailing scientific opinion.

Another New York/New Jersey Harbor Dredged Material Management Forum was also held in February. The PANYNJ brought out estimates that 56 percent of the Port's dredged material might be Category 3 and an additional 13 percent might be Category 2. Recall that prior to the testing changes approximately 95 percent was Category 1 and most of the remainder was at least Category 2. Now it was estimated that only about 30 percent of all of the Port's dredged material could go into the ocean without strict capping requirements, and even this estimate probably reflected a good measure of wishful thinking. This was partly because a large volume of sediments historically dredged from the Port came from Ambrose Channel, the entrance way into the Port from the Atlantic Ocean, and these sandy sediments had always been Category 1. So, although this new breakdown of categories was not news to the agencies, it really brought home to the public the realities of the situation.

A major discussion item at the forum was the planned construction and use of underwater pits dug into Newark Bay as a disposal alternative. One of the issues involved in this was who (and what kind of material) would have access to this disposal alternative. The PANYNJ was assuming the costs and supplying the effort for the task, but since the location was in New Jersey waters, would only New Jersey project materials be allowed to be disposed of there, given the limited capacity?

By March, there were still twelve applicants of the original forty or so, according to the Corps, that were sitting on their hands as far as testing because of the uncertainty of the legal climate. The Corps also had indicated that a major federal navigation project in the East River was being held up and funds allocated for it would probably have to be returned to their Corps headquarters. The problem was the Corps would have to test a third species and could not do so in time for its fourth-quarter funding award deadline. Somehow, however, a way was found, the project was tested and, eventually, its dredged material disposed of (more on this project later).

The National Marine Fisheries Service had by now completed a second study requested and funded by EPA. This one, similar to the earlier fish study, looked at levels of dioxin and other contaminants in lobsters collected from a wide area around the Mud Dump Site. This was done to address concerns that these valuable crustaceans could be more directly affected by contaminated sediments than the game fish species tested earlier, because of behavioral and food chain factors. The results showed that although the lobster muscle tissue (meat) was not appreciably contaminated, the same could not be said for the hepatopancreas (the "tomalley" discussed earlier). Some contaminants, including dioxin, accumulated in this organ to significantly higher levels than in the meat. Rhode Island was at this time coming out with an advisory against eating tomalley. New York and New Jersey were seriously considering doing the same.

For some time, word had been coming down from EPA headquarters that Clean Ocean Action was getting direct access to Vice President Gore's office and that an "understanding" was being negotiated. To staff in the agencies, it was surprising, and impressive, that a small local environmental group such as COA had actually gotten an audience at that level. There was also the feeling that the ball was now in a higher court, and little could be done except to see what would come of it. What eventually came was that a deal had been struck, and the details of it filtered down to the regional office. COA had negotiated an agreement with the vice president's office and EPA and Corps headquarters to phase out ocean disposal at the Mud Dump Site. The broad outline of the agreement was written in a letter signed by Vice President Gore and was somewhat further clarified in a memo from EPA headquarters. The agreement was finally described in more detail in a July 24, 1996, letter from the secretaries of the Department of the Army and the Department of Transportation and the administrator of EPA to New Jersey congressman Frank Pallone (and was thenceforth known as the three-party letter).

The four-page letter described the government's commitment to close the Mud Dump Site by September 1, 1997, and designate a surrounding area as the Historic Area Remediation Site (HARS), which encompassed the MDS. This entire area was to be remediated with "uncontaminated" Category 1 material. Among a number of other provisions, the letter stated that disposal of Category 2 material would be phased out by the September closure of the MDS and only "clean," Category 1 material would be allowed to be disposed of at the redesignated site after that time. It required EPA to evaluate the dumping criteria (most importantly, but not specifically stated, the bioaccumulation criteria for Category 1) over the next nine months. There were other provisions, including (1) the Corps was to schedule and expedite processing of the backlogged regulatory projects and ten priority federal projects; (2) EPA was to invest $1.2 million to support development of decontamination technologies for Port sediments and another $100,000 to facilitate pollution prevention in the Arthur Kill; and (3) disposal mounds at the Mud Dump Site could not be built higher than sixty-five feet below the surface. This was the government commitment.

The agreement also stated: "Most importantly, we expect that our commitments concerning the Mud Dump Site will diminish or eliminate the possibility of litigation challenging permits and the EPA rule change during the period prior to September 1, 1997. This proposal is predicated on that result." That, apparently, was the environmental groups' commitment. Although it left the matter of how "clean" material would be determined for a later time, this seemed like a reasonable agreement, with both the environmental groups and the agencies getting some of what they wanted.

There was also recognition in the letter about the commitment of the states of New York and New Jersey to implement the strategies of the agreement. This was accompanied by a stated allocation by Congress of $120 million for the PANYNJ, to be divided evenly between the two states to carry out the strategies.

9

CRISIS DEFLECTED

Now that permits would be processed again, the Corps had to do some serious work in planning and scheduling the sequence of projects to fit into the September 1997 deadline. This would be the last chance for Category 2 projects to qualify for ocean disposal. There were a number of projects that had a pretty good idea of their likely category. Projects that had done some preliminary testing and had failed the amphipod tests were looking elsewhere. Others that had conducted sediment chemistry had a pretty fair idea of what their bioaccumulation test results might be, within a good approximation. There is a simple equation for estimating potential bioaccumulation results from sediment chemistry and a few other data called Theoretical Bioaccumulation Potential. TBP had been developed some time before for this purpose, although it provided only an estimate, usually a conservative one. It was because of this last characteristic that, although the Green Book provided for regulatory decisions to be made using TBP rather than doing the full bioaccumulation testing, applicants usually opted for the testing even where they had the choice. The other data necessary for the formula were the sediment organic carbon content and tissue lipid content, as well as a default, experimentally derived accumulation factor.

If a project did not get into the ocean by the deadline, it might be a considerable time before the applicant could hope to get its areas dredged, which could in turn determine the continuing viability of its business. Every project and water-dependent business situation is different, but there is a limit to how long a lack of dredging can be sustained. Vessels can use a facility at high tide only, or they can be brought in after "lightering," whereby some of the cargo is off-loaded to another vessel while in deeper water. These methods add considerably to operational costs, however, and businesses cannot generally afford to operate this way for long periods. Many of the companies that had applications pending had already been operating this way for some time. So there were high stakes for these applicants to be included in this last-chance disposal window.

The Corps realized this and knew that it had to coordinate all aspects of permit processing beyond what had ever been necessary before. This included laying out a strict schedule for not only the sequencing of disposal operations but also for when each applicant must begin sampling and testing, and for how long a review period each agency would have, while also providing for an EPA QA/QC review period, a Corps' public notice period, and other aspects of the permitting and disposal processes. EPA conceded to the restricted schedule but requested, and got, agreement that EPA must consent on a project's categorization before the Corps issued any public notice. EPA correctly assumed that it would not get a lot of opposition from the Corps on this. Other regional EPA and Corps offices had after all been working together for some time under the new guidelines with this level of cooperation. Also, recently Corps headquarters had stepped in to "counsel" New York District when it appeared to be too recalcitrant on an issue and impeding progress.

One of these issues was the maximum height of disposal mounds that would be allowed at the ocean site. The "Gore Agreement" had stipulated that disposal mounds not be built to a height any closer than sixty-five feet from the surface. New York District did not want this clearance requirement to have to include the one-meter-thick cap. This would allow more material to be disposed of and greater flexibility in managing the site. EPA disagreed and believed that the sixty-five feet should include the cap. Corps headquarters sided with EPA, and that was the end of discussion on that issue.

Some of the projects that were on line for the deadline rush had not done any testing recently, as had the partially "grandfathered" projects, and therefore had the privilege of being the first ones to perform all the new testing in the revised regional manual. As has been noted, the major testing changes that could result in a failure or a higher category level were either the amphipod tests or the enhanced bioaccumulation tests. The states of New York and New Jersey were also becoming more active players in the testing now. They had both drafted dredging evaluation manuals of their own and had realized that the required analysis of alternatives to ocean disposal would place more responsibility on them than ocean disposal ever had. The Gore Agreement also required that one of the hoops applicants had to jump through to get into the one-year disposal window was to secure acknowledgment by the relevant state that there were no land-based alternatives available. Since most of the critical projects with the longer processing delays were in New Jersey and the ocean disposal site was off the Jersey shore, public officials there felt the heat sooner and had started to address the issues in a programmatic way sooner than New York did.

Both states had been planning and setting aside funds to begin to seriously address the dredging issues. They were now participating directly with the federal agencies in devising sampling and testing plans on individual projects. The first project that was handled by this new approach was the Naval Weapons Station Earle, in Sandy Hook, New Jersey. The New Jersey Department of Environmental Protection wanted considerably more analyses than had been proposed by EPA and the Corps. This could be a problem if the differences were not worked out, since the state could withhold the water quality certification necessary for the Corps to issue a permit.

The state believed additional analysis was needed to enable data evaluation against its upland disposal criteria, in case an upland disposal site might end up being needed for the material (i.e., if it was Category 3 or Category 2 and did not make the disposal window). Disposal of material on land sites typically requires different analyses than for ocean disposal. For instance, one type of test is designed to represent conditions that could occur in a landfill if contaminants were to leach from dredged material into groundwater by way of rainfall and percolation. It is called Toxicity Characteristic Leaching Procedure and is a standard test for hazardous material. Variations on this test have been developed for dredged material, and all such tests use only chemical methods; no bioassays are involved.

The Navy representatives were becoming concerned that the State and federal agencies were at loggerheads, which would put their facility's dredging and upgrading schedule in jeopardy. There was no reason, however, that sampling requirements for the different kinds of tests could not be reasonably combined if both parties would take a wider view. During a meeting held on this issue, the Navy people left the room for the discussion of technical issues to be continued with the federal and NJDEP staff scientists. When the navy people were brought back in at the end of the discussion, they were so relieved that agreement had been reached that they hardly cared to know what the details were (which was uncharacteristic of them).

Exxon (Petrochemical Terminal in the Arthur Kill) had already done some of the additional testing that had been anticipated and did not want to go out and resample for more tests. It petitioned to be classified a Category 2 project on the basis of the testing done, thus hoping to obviate additional tests since its samples had already passed the toxicity tests. This was akin to plea bargaining by admitting to a lesser crime before any charges are even brought. EPA decided it could not be a party to this, since however slight the chance, the additional tests could conceivably result in a finding of Category 3. Normally the toxicity tests alone could fairly confidently identify a Category 3 project. It would not be expected that a project could pass toxicity tests and have bioaccumulations so high that it would have to be considered worse than a Category 2. But since Exxon's toxicity test results were very much on the borderline, a conservative decision was believed to be warranted, and Exxon was told it would have to do the same additional testing as the others to qualify for the disposal window.

By October 1996 there were sixteen active private applications, plus another twelve federal navigation projects pending for ocean disposal. Corps of Engineers headquarters had scheduled a meeting with EPA and Corps regional staff. It wanted to make clear the importance of getting materials from these projects tested, approved, and disposed of by the deadline in a smooth and orderly fashion. It was pretty clear that the Department of the Army wanted to make sure there would be no screwups, on the Corps' end at least, in implementing the White House–brokered agreement. A Corps headquarters representative at the meeting made clear to the New York District that he was not satisfied with its process for scheduling and tracking applications. An even more regimented tracking method was developed for the District to follow.

This kind of internal criticism was unusual in an open meeting with other agency's personnel in attendance (in this case EPA's), especially since it appeared to some at the meeting—including the author—that the District's process left little room for improvement as presented. The representative also emphasized his view, in so many words, that whatever the relationship between the agencies had been in the past, or perhaps would be in the future, for right now and until the end of this process they had better be very much like two cogs in a well-oiled machine.

In early October also, the first meeting was held with the environmental groups since the Gore Agreement. The subject of the meeting was the bioaccumulation thresholds used to distinguish between Categories 1 and 2. EPA proposed using the human health risk–based values developed by the Criteria Workgroup's bioaccumulation subgroup. Clean Ocean Action balked at this, not because it believed the values were not protective but because the procedure used to develop them did not take into account protection for ecological resources. That is, although using these thresholds might be protective for the ultimate consumer in the food chain, humans, some lower thresholds might be necessary to protect sensitive species within the food chain—or in another food chain.

This argument was not an insubstantial one from a process standpoint, but EPA staff believed from preliminary work already done that the proposed values would also be reasonably protective against ecological effects. It was true that ecological thresholds had been estimated to be lower than human health thresholds for some contaminants under some conditions. For example, a larval or juvenile fish that accumulates a certain concentration of a contaminant, say PCB, through its diet may experience effects such as an inability to produce viable eggs or to grow adequately. That same concentration in a human may have no discernable effects at all, for reasons such as an ability to sequester the contaminant in fat reserves or because the biochemical toxic action mechanism is more tolerant. The ecological thresholds that had been estimated by the bioaccumulation workgroup were in rudimentary stages that left much room for subjective argument on many points. To narrow down the range of uncertainty for these points of contention would require the assembly and consideration of more information than was readily available.

Nevertheless, the environmental groups strongly suggested that thresholds be developed to address ecological, not just human, endpoints. Still reeling from its recent court loss, EPA headquarters appeared willing to accede to the groups' demands to the greatest extent possible. The nine-month window for EPA to "review the ocean disposal testing requirements and ensure that any further revision reflects both sound policy and sound science," in the language of the Gore Agreement, was hanging over EPA headquarters. The political game board had clearly been set, and the side that was flush with victory was going to get its way.

As was noted, the main problem in developing ecological-based bioaccumulation thresholds was that there was precious little relevant or good information available. The human health-based values that the bioaccumulation workgroup proposed were based on data from peer-reviewed, published scientific studies and had been developed through widely accepted risk assessment methods. The group had been working on ecological thresholds also but had found little relevant supporting data, and the uncertainties in the risk methods left considerable room for objective argument. There were a variety of ar-

ticles in the scientific literature on various studies that involved bioaccumulation, and ecological values could theoretically be calculated based on some
of the results. But the studies used a hodgepodge of different methods and
experimental conditions, and for the most part did not lend themselves to appropriate application for the desired purpose. Further, there had been differing views among the workgroup members on assigning appropriate values to
safety factors, exposure factors, species differences, and other variables in the
equations of the risk-based approach. To try to develop threshold values that
would be scientifically defensible would take considerably more time and resources. This was the problem that the workgroup had run up against before
and could not surmount in the time it had. Nevertheless, this was apparently
what was required now, and the agencies now put their full effort into it.

First, a technical meeting was held in the EPA regional offices that brought
together scientists from the Corps' Waterways Experiment Station, EPA's Nar-
raganasett lab, and other government offices. The main subject of the meeting
was to try to come up with bioaccumulation thresholds, or at least an approach, for evaluating PAHs (polynuclear aromatic hydrocarbons) in the bioaccumulation test results. These are a group of organic contaminants that are
typically generated from a variety of man-made and natural sources including
automotive exhausts and other air pollution sources, petroleum product spills,
or forest fires and volcanic eruptions, and are often found in high concentrations in Port sediments. Several of the PAHs are known human carcinogens.
The environmental groups had targeted this group of contaminants based on
test results of several recent projects that had fairly high concentrations, so
PAHs were given the first round of attention.

To get all these experts together on relatively short notice for even a day
had not been easy, so there was an intense effort to try to reach consensus before the end of the day. A significant body of scientific literature that had been
assembled and distributed prior to the meeting was summarized and discussed.
At first the discussions were fairly wide-ranging and unfocused, but by the end
of the day, after reviewing the literature articles and several approaches that
were brought for consideration, the participants actually hammered out some
general threshold values for PAHs. Corps and EPA staff, including some from
headquarters, were tasked to flesh out the rationale and supporting documentation. Then there was a rush for the airports.

The question that some of the involved scientists had after the dust had
settled was, were these procedures and threshold values actually good ones?

The process had certainly been quick, and it left some scientists wondering how much the political pressure to get something out had overwhelmed deliberate scientific evaluation. There were some continuing discussions for a time among some EPA scientists about other information that may have received less than full consideration and that could have resulted in lower thresholds, but the ball was very much rolling by this point. The meeting set the stage for a continued frenzy of work to develop ecological thresholds for the rest of the contaminants.

EPA headquarters and a New York regional scientist (author) traveled to WES in Vicksburg, Mississippi, to concentrate on assessing the literature and other information with Corps scientists in developing more thresholds and to draft a decision memorandum. This decision or authorization memo documented all the scientific support in the literature for development of the thresholds and would come to be known as the Supermemo. It included evaluation methods for all the tests and the threshold values for all the bioaccumulation contaminants.

It was the largest decision memo ever written for a dredging project, with pages of data tables and written documentation supporting the bioaccumulation thresholds. It was first prepared for the East River federal navigation project by the bi-agency team working at WES. As mentioned previously, the project was running very short of time before a deadline expired for it to receive the allocated federal funding. The Corps could not authorize it until it had EPA concurrence, and the Vicksburg team from both agencies was working on the authorization memo. It was completed at about 2:00 a.m. on December 2, 1996, so that an authorization process necessary for the project to be funded could start that morning. The memo would be used in "boilerplate" form for all of the following projects in the race for the September 1997 Category 2 deadline. It continued to be used for dredging projects thereafter, including in Region 1 EPA in Boston, which began using it as a model for its ocean disposal evaluations in 1999. Then a contingent from EPA headquarters moved back to the region to continue the work and to help fine-tune the decision memo. Headquarters was not satisfied on certain technical and legal issues, and there was continuing rewriting for another couple of weeks getting it ready for the next string of projects.

After the East River decision, the rest of the projects that had been anointed for this last-gasp disposal window were in various stages of being sampled and tested, and their results quality-assessed, by the regional EPA lab in Edison,

New Jersey. Then, decision memos were written up in a continuing flurry of activity. More staff were again brought into the dredging program from other branches in the regional office to help conduct all the technical reviews. The ocean dumping coordinator from EPA Region 1, Dave Tomey, was asked to come in on a one-week detail to help review data and write some of the decision memos.

Disposal operations for the remaining certified Category 2 projects continued without stop (and without much incident) until the deadline on September 1, 1997. The critical dredging projects for the Port's navigational channels and shipping berths that had been backlogged for so long were now finally dredged. The Port shipping and waterfront industries were greatly relieved to see that real action was being taken and that the crisis appeared to be stemmed. It was as if a dark veil that had hovered over the future of the Port had been lifted.

However, as discussed further below, the Port's dredging and shipping problems were far from being solved over the longer term. It would have been a very good bet, though, that had not the Gore Agreement been concluded and implemented as successfully as it was in unclogging navigational arteries and berthing areas, the Port of New York and New Jersey would be a very different and diminished version of what exists today. The Port will continue to face problems in the dredging and shipping industries and also has critical needs for other improvements to the transportation infrastructure such as roads and rail facilities. For all its continuing problems, the Port is very much a vibrant and growing centerpiece for an area that is widely considered the enterprise capital of the world.

Then, the Mud Dump Site was de-designated at the same time that the new Historic Area Remediation Site (HARS) was designated. (The author was not there to witness the grand conversion, having left EPA in late June to start a restaurant business on the Jersey shore. Any naïve ideas that this change might provide some respite to the psyche from the previous sustained periods of high stress "critical" work on the Port dredging program would soon be rudely dislodged. In any case, he was back in New York two years later working for the NYSDEC in its Marine Resources Bureau, dealing with, of course, Port dredging issues.)

10

LESSONS

Before going on to the present situation and the possible future for the Port, it may be instructive now to look at what potential lessons may be learned from the missteps that caused and sustained the dredging crisis. It must first be acknowledged that looking back now is obviously with the benefit of hindsight. Nevertheless, although the mistakes made can be largely attributed to human nature, an objective look back might identify lessons that can be applied to other environmental problems and issues.

Since it was the testing changes that brought on the need to decide how to proceed with the program, their implementation was the most direct trigger for the Port dredging crisis. The Corps and the PANYNJ never denied the need for some testing changes, though the changes would almost certainly be in conflict with their main missions. There was plenty of criticism and jousting over particular items, but none of the professionals that were involved in the process pronounced that testing changes should not be implemented. Although in their internal meetings criticism of the revised testing may have been somewhat more expansive, any general condemnation would surely have gotten short shrift by management.

The New York District Corps of Engineers itself would probably not deny at least some responsibility for the crisis, if for no other reason than its preeminent role in Port dredging. It would be difficult to imagine that the Corps has not already undergone some intense self-analysis, at some level, as a result of this experience. A conclusion that might have been reached is that there must be more flexibility within the system when faced with new kinds of challenges. The Corps' hunker-down, circle-the-wagons type of reaction to the potential threat that ocean disposal could be severely restricted did not work. It only slowed the inevitable outcome, without buying time for an alternate plan to be put into effect. Alternatives were investigated, but most were found lacking for a variety of reasons, and the one that most effort was expended on (borrow pits) turned out not to have been a good choice. The fact that alternatives to ocean disposal of the kind initially rejected by the Corps have been implemented since the crisis, including New Jersey land-based applications, demonstrates that they were in fact possible.

In hindsight, it may be easy to blame the Corps for its perhaps less than robust attempts at developing and sponsoring alternatives. It should be kept in mind, though, that some of the alternatives that had been identified early on had problems not only from a cost standpoint but also because they would have involved public/private partnerships for land acquisition and operational management. The New York District dredging program did not have experience in this area. The District Corps had always managed its federal navigation projects as well as its regulatory permit program as a wholly federal responsibility. Also, some alternatives that were identified had fairly negative public acceptability issues, mainly related to NIMBY (not in my backyard). Any new plans for disposal or treatment of contaminated material would have been skeptically received in some areas by the local environmentally minded public. And proposals by the New York District Corps to dispose of the Port's dredged material, which was increasingly being publicized as a toxic threat, could have been assumed to be greeted with outright opposition. But the Corps also had the responsibility to fully follow through on any technically and environmentally feasible options, after taking into consideration the likelihood of local opposition.

Public opposition did not turn out to be a big factor in the New Jersey land application projects. In fact, some of the potential areas identified in the Corps' early investigations included areas in the vicinity of some that New Jersey eventually implemented. In those investigations, the Corps was looking for vacant, government-purchasable or leasable land that could be used for dredged material dewatering and placement. Instead, the state of New Jersey used shared public-private means to implement its land-based alternatives. Issues relating to NIMBY were not a huge hindrance, because most of the areas in question were zoned industrial. The question then is, why could it not have been done earlier rather than later?

There have certainly been occasions during the course of human affairs when a difficult or costly decision presented itself but the driving force for implementing it had not yet built up sufficiently (e.g., cleaning our rivers, global warming?). When the pressure became high enough, be it public opinion or whatever other driving force, the decision was then deemed sufficiently justifiable. Once the Port dredging crisis ensued, the public opinion and economic pressures were certainly sufficient to justify increased disposal costs. To what degree the Corps' failed response to this challenge can be attributed to the above "ripeness" factor, or organizational inertia, or some other factor, ultimately has to be left for the reader to ponder. To this observer, it appeared that the effort the Corps put toward alternate disposal options may have been marginalized by the belief that the Port's dredging, and thereby its viability, would never be allowed to be put in jeopardy. If there is any truth to that observation, the lesson for decision-makers here might be to not bet the farm on assumptions like that.

EPA also obviously had a role in the alternatives issue, from its initial commenting on the Corps' early investigations and on into the unfolding events. Region 2 initially thought it could leave all responsibility on alternatives identification and development to the Corps, since dredging was mostly a Corps program. EPA may have been correct from a strictly programmatic standpoint, but perhaps a little more foresight would have revealed that the testing changes that *it* was more forcefully advocating would result in some wrenching changes in the Port dredging situation (as was detailed in the memo discussed in chapter 8). Also, as was discussed earlier in chapter 5, a Corps Planning Division executive once proposed a large combined disposal scheme, which was politely ignored by EPA. This particular proposal may have been too much of a paradigm shift for EPA to get a grip on, but there was nevertheless little concern at an early enough stage to work more seriously with the Corps (even if the Corps was not asking) toward the development of real alternatives to ocean disposal. After all, EPA had just prior to and during this period taken a decisive role in developing nonocean alternatives to the disposal of sewage sludge in the region.

Region 2 EPA's decision not to directly address the environmental community's concerns regarding the Port Newark/Elizabeth project may well have resulted in a missed opportunity to resolve the scientific issues and preclude the lawsuit. It is of course conjecture and hindsight to presume that this could actually have occurred, but it is worthwhile to consider another direction that could have been taken. The major issue for Clean Ocean Action and the other groups was how the agencies were assessing risks from bioaccumulation of dioxin and other contaminants, mainly dioxin at that juncture. The Criteria

Workgroup's bioaccumulation subgroup was in place at the time and working on bioaccumulation thresholds. It was not focusing on the dioxin thresholds, since EPA had strongly suggested that the existing management plan it had worked hard to negotiate with the Corps be left alone, at least for the time being. EPA felt that it would be a better use of time to develop thresholds for contaminants that did not have appropriate risk-based thresholds.

Although this seemed to make sense and be fair from EPA's standpoint, it did not address the environmental groups' concerns with respect to the Port Newark/Elizabeth project. Their concern about this project, including its potential as a precedent, involved mainly the dioxin issue. If this issue had been addressed head-on with maximum resources (as was later done, starting with PAHs), things might have played out differently.

This relates back to the way that the workgroups had their tasks identified and the way they were staffed and managed. The dredging forum created workgroups with representation from every possible interest group in the Port. This followed EPA's national trend of using inclusive partnerships and regulatory negotiation with stakeholders, as in its Common Sense Initiative and project XL programs (U.S. EPA website). Oversight of the workgroups was by a committee of the workgroup chairs and higher-level managers of the federal and state agencies, called the Dredged Material Management Integration Workgroup (DMMIWG, commonly known as the Dimwigs). This approach was seen as the ultimately fair and modern way to solving the Port's environmental problem.

One problem with this approach was that workgroups often had members who had no technical or scientific basis for being there. Some of these members would have been the first to admit that that was the case, but they were appointed by their agency or group and felt obligated to attend (many attended once or twice but did not continue). Some of these "nontechnical" members did want to be there to attempt to influence the proceedings as best they could to promote the interests of their group. People do not have to be familiar with or learn toxicological methods, for example, to know that if they would like a less stringent threshold, pressing for a less stringent trophic transfer factor or a higher toxicity limit would have that result. The majority of the discussion and all the investigative work ended up being done by the few "technical" members of the group. And some of these could spare only a limited amount of time for the effort. This was the situation, at least, with the larger Criteria Workgroup. Regulatory reinvention, reg-neg, or any of the other new inclusive approaches were not appropriate for these complex technical issues, at least in the way they were being practiced in the dredging forum. It was not an efficient process.

It would have been better to let the Dimwigs be the only residence for the policy-level people and to designate only technical experts as members of the workgroups (as was the case essentially for the bioaccumulation subgroup). With those designations should have come time commitments sufficient to complete the effort, supported by the group members' sponsoring organizations, so that results could realistically be achieved. Then, a very focused approach could have been taken first to ensure that the problem was well defined, coordinating this with the Dimwigs, then to address it with a concerted, dedicated effort. In the case of the bioaccumulation subgroup, allowing for the full commitment of its members and providing the necessary resources may have resulted in agreement on dioxin. As the leader of the forum, EPA could have directed a concerted effort to resolving the core of the dioxin issues and pressed the other forum partners to follow suit. The 10 pptr dioxin threshold was already fairly low, and because of the low-concentration measurement problems, a good case could have been made that there was not much lower it could reasonably go. Focusing from there on disposal management issues for dioxin might have resulted in a consensus.

To complicate matters further on the EPA side regarding its general negotiating position on testing revisions, reshuffling of management occurred a couple of times between programs in the Water Division, one of which was the dredging/ocean disposal program. This was standard procedure in EPA to give managers a broader experience over a number of different programs. It may be an appropriate management tool generally, but the dredging program at this time required a consistent focus and approach. There was certainly no such shifting on the Corps side; the Corps was consistently focused. Following one of these shifts in January 1992, a "fresh" negotiating approach undoubtedly resulted in unnecessary delay in reaching agreement on key issues. These attempts resulted in backsliding from movement that had been going in the right direction, and in both cases the fresh starters were disappointed after several negotiating sessions with the Corps and soon got back on course. That course was to recognize the irresistible factors that were on EPA's side, that the revisions were scientifically the right thing, and that there was a general consensus about all this in both agencies.

The back-and-forth shift of the EPA dioxin bioaccumulation threshold between 10 pptr and 25 pptr did little to dispel the notion held by some in the

Corps and elsewhere regarding EPA's ambivalence on key issues. It may have been true that EPA was subject to perhaps a wider range of pressures than the Corps (which felt beholden mainly to its dredging applicants and its own dredging needs) and had to perhaps more seriously balance exhortations received from both the environmental and the dredging communities. There was, in addition, political support by various members of Congress on both sides of the issue.

The strategy should nevertheless have been to develop supportable science (which EPA did) and then hold solidly to a position on the key issues. Instead, the Corps had been holding more firmly and by so doing may have thought it would prevail in the end. Yet, every one of the contested procedures or issues, from amphipod and dioxin testing, to increased early information sharing on permit applications, to conducting chemical analysis of project sediments, to establishing (interim) bioaccumulation criteria, was eventually implemented when each was again seriously pursued. Whether, and to what degree, EPA's waffling delayed the eventual breaking of the project logjam can only be surmised. It may realistically have had little effect, in the overall scheme of things, but it may have had some, and the lesson here may be that the old adage about persistence being the key to success may well have worked better for EPA than it did for the Corps.

The PANYNJ was the other big player in the Port's dredging crisis. Recall that in a Save Our Port meeting in 1989 the Port Authority had the opportunity to take a better look at alternatives to ocean disposal but let it pass. The Port Authority later came under different management both at middle and top levels, and there appeared to be a little more flexibility presented afterward. At the dredging forum meetings the Port Authority held to positions consistent with the Corps, as might be expected. Yet, there were signs at that point that it may have been more willing than the Corps to try other approaches. For example, as cochair of the DMMIWGs, the PANYNJ's representative and dredging program manager Tom Wakeman worked seriously with the environmental interests and other groups to address the issues. The other cochair of the group is Jim Tripp, an attorney at the Environmental Defense Fund, one of the environmental groups involved in the Port dredging issues. The DMMIWG still meets at the time of this writing but is no longer directly engaged in the central issues being dealt with by EPA and the Corps, such as the bioacumulation criteria effort. It has been dealing mainly with broader issues such as Port efficiency improvements and how they relate to the shipping industry and the regional transportation infrastructure, as well as holistic approaches for environmental remediation in the Harbor.

The PANYNJ has since then taken steps to back its stated commitment to adopt a more comprehensive approach in dealing with the Port's dredging issues, which included more consideration and support for environmental components. It has been instrumental in seeking and obtaining congressional allocation as well as providing its own funding for investigating environmental mitigation options as part of comprehensive long term plans for the Port. The plans include deepening of some channels and berths, and structural waterfront improvement for piers and terminals. The mitigation includes restoration and improvement for contaminated sediment areas or impacted wetlands around the Harbor.

This brings back an interesting possibility regarding another path that might have been taken but that can only be surmised from hindsight. One example of another approach occurred in the port of Baltimore, where the port authority took a proactive position with the Baltimore District Corps in the construction and management of Hart-Miller Island, a large containment complex created in the 1980s to manage contaminated dredged material from that port. Such a development would have stood little chance within the Port of New York, as there were early and insistent warnings against any proposal for establishing containment islands in the harbor, mainly because of the loss of marine habitat that a containment island would have entailed. The approval for Hart-Miller was probably due mostly to the fact that Chesapeake Bay encompasses a considerably larger area than New York Harbor and thus the containment facility's construction resulted in a proportionally much smaller loss of habitat. However, one might wonder what could have been accomplished if PANYNJ had taken a comparatively proactive stance at an early enough stage (i.e., during the Corps' alternative investigations before 1990). For example, it might have taken it upon itself to build a New Jersey land-based alternative to ocean disposal, or the Newark Bay Contained Disposal Facility (discussed later), before the crisis could really set in.

No doubt the PANYNJ would still have had to work closely with the Corps to accomplish such things. At the time, it was apparently the New York District Corps that set the stage and tone regarding developing ocean disposal alternatives, and, as was earlier inferred, the tone was not overly positive. (The author recalls receiving a phone call in the early 1990s from a dredging applicant voicing a last-ditch request whether any disposal alternative at all could be suggested. When told that, regrettably, the only suggestion was to perhaps try working a land-based solution through partnerships or other arrangements, he indicated this had in fact been discussed with a midlevel Corps manager, who told him it would probably be a waste of time.) It is not clear how broadly this view was held in the District Corps, or to what degree it may have influenced the PANYNJ regarding going forward or not with alternatives under its own steam. In any case, it is clear the PANYNJ was closely linked to the Corps regarding Port dredging, and at least its early reactions can be viewed as having been driven from the same mission orientation and philosophical view. The lesson here again is to strive for the greatest flexibility in meeting new challenges, and perhaps for upper management to look more critically at entrenched positions taken by operational managers.

Lastly, the environmental groups played a major role, perhaps not so much in the development of the crisis but certainly in how it was handled. The pressure that was felt by EPA (somewhat less by the Corps) from these groups was a big factor in the decisions that were made in addressing the crisis. EPA certainly had the obligation of implementing updated testing requirements, regardless of any pressures it may have felt from the environmental groups to do so. Whether, or to what extent, COA's and others' hyping the dangers of the Port's sediments may have increased the difficulty of finding acceptable land-based sites is not clear. It's not likely, in any case, that this factor was a very significant one.

In a previous chapter of this book there was some criticism of "the end justifies the means" type of approach used by Clean Ocean Action in gaining the Jersey shore public's interest in and response to ocean disposal issues. This local success certainly, however, ended up gaining COA access to the New Jersey governor's office and eventually to the White House Council of Environmental Quality and Vice President Gore's office. This access in turn resulted in the agreement to ban Category 2 material from the ocean. Even though the ban was actually pronounced earlier in a less official sense by the EPA regional administrator's office, that too could be said to be the result of earlier public advocacy successes and pressure by COA.

On December 11, 1995, after his initial decision in favor of the government was overturned by the Third Circuit Court of Appeals, Judge Debevoise pronounced that the permit issued for the Port Newark/Elizabeth project was not lawful, because not all regulatory requirements had been met. Clean Ocean Action and the other environmental groups claimed a great victory, and they were able to get a lot of mileage out of this in their subsequent negotiations with the administration and the federal agencies. The findings of the decision were not, however, used to stop any project, and materials from other projects were subsequently disposed of with less testing than was done for Port Newark/ Elizabeth. EPA modified the Ocean Dumping Regulations to remove the requirement for performing bioaccumulation tests on the suspended phase, and later to remove the requirement for using three species in the solid-phase tests. The most direct result of the environmental groups' victory, then, was that it forced the changing of the regulations to more closely follow what the agencies had wanted to do all along. The judge's final ruling was therefore essentially a one-time decision on a permit for sediments that had already been dredged and disposed of, and a hand slap to the agencies based on what must be considered to be technicalities.

They were technicalities in the real sense that the environmental groups did not object to the main regulatory changes removing the testing requirements that they had sued to enforce! An argument could be made that the groups' demand for strict adherence to the letter of the regulations was used mainly for the purpose of winning the case, not because the groups seriously believed that the procedures in contention actually provided for more environmental protection. If they truly thought that three benthic test organisms, for example, were necessary to adequately evaluate sediments, why did they not object to the proposed regulatory changes to eliminate the requirement? Admittedly, this is probably a naïve view of how our legal system works. This was not the first time, nor will it be the last, when similar methods were used to effect desired changes in government agencies by whatever legal avenue was available. Also, the environmental groups probably did not initially intend for the lawsuit to result merely in some regulatory changes but rather to bring about more conservative assessments for ocean disposal. That end was in fact achieved to some degree, and probably to no small measure because of the lawsuit. Still, it leaves a bit of a bad taste when a legal end run is played around the obstacles of a difficult issue, especially when the main obstacle remains in place. The obstacle in this respect is the apparent inability of the agencies and environmental community to confront a difficult scientific question and deal with it head-on.

To clarify the last point, we need to look first at the understanding and treatment of science in the courts and in the environmental community. It was apparent that neither Judge Debevoise nor the appeals court panel understood very well the underlying science behind the regulations. The "plain meaning" of a regulation as viewed from a legal standpoint may be quite different from the original intent of the law and regulations if that intent is clouded by awkward or unclear language (as existed in this case). Another possibility of course is that the original intent of Congress itself could have been cloudy. Yet the court had to rule on complex technical questions that were in many cases subjective and difficult to fully grasp even for the experts who routinely dealt with them. Undoubtedly the judges could confer with other experts, and it was rumored that Debevoise had a scientific staff member that worked on the issues. The appeals court panel decision gave no evidence of having considered the scientific merits to any degree at all. So is it realistic to think that judges totally unfamiliar with the underlying science can appropriately decide technical questions of this nature, even if they might have the advice of persons with a general scientific background? This observer thinks not.

Unfortunately, environmental and scientific questions will no doubt arise many times in areas where such a standard legal approach is taken. One can only hope that most of these questions will be on issues that can at least be somewhat more clearly defined, and amenable to broader scientific expertise. In cases such as this one, however, there should be better provision for courts to be able to set up a panel of real and independent experts in the field and then abide by their decisions. These kinds of panels or expert groups are nowadays assembled to address major national issues such as global warming or stem cell research policy. It seems to this observer that the merits of this case deserved consideration by real scientific experts, who could have come from academia or other institutions independent of the government agencies.

It was also obvious that the environmental groups were laboring under a less than full understanding of the science underlying the regulations and the testing manuals. Without making a federal case of this and in the interest of brevity, an example comes to mind. In a meeting following the environmental groups' court victory, it was agreed that the federal agencies would review the scientific literature to try to develop ecological-based (and not just human health–based) bioaccumulation thresholds for PAHs. The environmental groups volunteered to supply a number of scientific articles that they believed demonstrated the availability and applicability of data for this effort. True to promise, a bibliographic listing of articles with their citations was received from the groups, and then the entire articles were duly retrieved and reviewed by agency staff. As it turned out, most of the articles were not even nearly relevant to what was needed for criteria development. It appeared that just a keyword search for PAHs was done and the output printed and sent in—it was doubtful that any of the articles had actually been read.

The point here is not to ridicule the groups for their scientific shortcomings. It would not necessarily be fair to expect that Clean Ocean Action and perhaps some other groups involved should have the relevant scientific experts on their permanent staff, but they could have done better in reaching out to the larger groups that did have them. The groups did have occasional assistance from two scientists during the crisis period but this did not meet the needs of the situation. COA and its co-complainants spearheaded a major case with significant national import and a wide following. There could and should have been a better effort to retain qualified experts from the national environmental groups, such as the Environmental Defense Fund or the Natural Resources Defense Council. And the major groups could have been more helpful if they had committed experts to participate over the long term in serious scientific discussion.

Sarah Clark, a staff scientist with EDF, wrote a letter in 1992 to the FDA requesting that it take a stand on the fact that the FDA 25 pptr level for dioxin in the Green Book was not an FDA action level, and that it inform the Corps and EPA of this. Ms. Clark had previous to this provided input on the dredging issues but would soon leave the region to start her own consulting firm. Much later, in February 2000, Dr. Kristen Milligan testified for Clean Ocean Action before the congressional Subcommittee on Fisheries Conservation, Wildlife and Oceans. The testimony criticized in some technical detail the bioaccumulation thresholds being used for placement of dredged material at the HARS. Dr. Milligan had also done some previous work for COA in the review and commenting on dredged material projects and protocol development. But in some of the crucial periods discussed earlier, there was no serious participation by scientists from any of the groups. A little more serious effort on the part of the environmental community as a whole to provide consistent scientific participation could have gone a long way.

Serious participation in the bioaccumulation subgroup by a qualified representative could perhaps have led to more real changes than were otherwise obtained. A possible example of how this could have occurred was in the evaluation of dioxin bioaccumulation for Port dredging projects. As noted previously, the agencies had hammered out a dioxin protocol prior to the dredging forum workgroups being established, and EPA strongly desired the Criteria

Workgroup to focus on the other contaminants that did not have updated threshold limits. During that time there had emerged a broad scientific consensus that coplanar PCBs, a small subset of that group of compounds, act very much like dioxin. Further, they had been assigned their own set of toxicity equivalents, similarly to how equivalents for the different congeners of dioxin were assigned.

This would mean that, in adopting the consensus view that these PCBs be treated like dioxins, the bioaccumulation results for these PCBs would be added to the dioxin total equivalents for a project. The end result might of course have been that more projects would have failed the dioxin protocol. (A limited evaluation of this possibility by the author some time ago indicated that the difference in failure rate, because of various technical factors, would probably not have been large.) Whether that would have been the case or not, it cannot be denied that a better and more inclusive understanding of the potential effects of Port sediments on the marine environment would have been obtained. A more committed and expert involvement by the environmental community in the workgroup (along with some flexibility by EPA) might have helped bring such an improvement to bear.

Perhaps the most important difference that might have resulted from more active expert participation by the environmental groups, however, was that more protective bioaccumulation thresholds generally might have been established prior to the lawsuit. Although the blame for this can be shared by EPA and to some degree the Corps, an opportunity was probably lost to establish more widely accepted thresholds because of the lack of active expert participation by the environmental groups during the crucial period of the bioaccumulation subgroup's efforts. It is these thresholds that in fact remain subjects of continuing complaint.

11

THE PORT TODAY AND TOMORROW

Despite the belief of some in the Port community, the dredging crisis may not be over. For smaller waterfront users, such as marinas and smaller terminal facilities that cannot afford the high cost of testing for ocean disposal or pay the greatly higher costs of alternative methods, the situation is still critical. Furthermore, the dredging cycles for some of the larger projects that were squeezed in under the September 1997 ocean disposal deadline are coming around again. Materials from those and other Category 2 and 3 projects do not have an ocean resting place in their future. Also, the critical focus being placed on the bioaccumulation thresholds could result in downgrading for some projects that had previously passed as Category 1, making their sediments unacceptable for remediation material in the ocean. (It is also possible of course, if unlikely, that bioaccumulation thresholds could be made less restrictive than they are now.)

Many of the dredging areas or projects with sediments that had historically been disposed of in the ocean have been set aside or have had to find alternative disposal or management methods. A number of water and land-based disposal alternatives have been developed and used since the closing of the old Mud Dump Site and designation of the HARS. Disposal pits in Newark Bay were constructed and used for local dredging projects by the PANYNJ. Several land disposal alternatives were applied in New Jersey, including remediation for closure of landfills and as base material for a mall parking lot. The land disposal alternatives were and typically are quite a bit more expensive than the old ocean disposal practices.

The Newark Bay pits were the first project of their kind in the Port, and a lot of planning, funding, and permitting effort went into developing them.

This had started during 1996, and in November 1997 the PANYNJ completed construction of the first cell of the Newark Bay Confined Disposal Facility. These pits would have a limited capacity, about a million cubic yards, and were used mainly for high-priority Corps federal and PANYNJ projects. The planning that had to be done for the sampling and evaluation necessary for permitting this project was not a trivial task. Deep pits were to be dug in the shallows just offshore from Port Newark/Elizabeth (between the existing navigation channels), an area containing sediments ranging (with depth) from recently deposited to heavy industrial to preindustrial. Sampling cores had to be sectioned appropriately for testing to reasonably describe these layers, because disposal alternatives would likely have to be used for the materials. The heavily contaminated industrial layers had to be handled differently from the preindustrial layers (which were clean and had potential use for beneficial purposes such as construction aggregate). Since only the cleaner material, mainly the preindustrial layers, could be placed into the ocean, it was critical that these boundaries be purposefully defined because the pit construction depended on it.

A major issue was that most of the contaminated industrial-age layers would have to be dredged and stockpiled, then placed back into the bottom of the pit. Since the pits were designed for contaminated material, it made sense to use some of their capacity for disposal of the "problem" material excavated for their construction. This meant that construction plans, including the pits' final dimensions, relied on a predetermined volume necessary to contain material that would need to go back in before other project materials could be placed into them.

From the preliminary analyses, though, the layers were not so clearly defined, and the layer-specific sediment chemistry and other test results were negotiated over at some length. The matter was finally settled, mainly between EPA and PANYNJ staff, and once the permitting was done, the actual implementation was fairly straightforward. A channel had to be dug from one of the navigation channels in Newark Bay for the construction tugs and barges to operate. There were no environmental or operational problems reported in construction, nor later in placement of dredged materials. The state of New Jersey had initially decided that only project sediments from New Jersey waters could be disposed of there, but some limited use by dredging projects in New York waters was eventually allowed. The Newark Bay Confined Disposal Facility was one of the first low-tech alternatives to ocean disposal that proved to be successful.

There have been other alternative methods found for some projects, and further efforts may identify and implement more land-based applications, even including some decontamination technologies. New York State has recently applied techniques developed at New Jersey landfills and remediation sites at two smallish landfills along the Jamaica Bay shoreline in Queens. There is also movement toward authorizing the large capacity at the huge Fresh Kills landfill on Staten Island for dredge disposal. However, these options all require some form of treatment for the material, typically solidification/stabilization. Although this is among the least costly of the treatment technologies, the costs for such operations through New York City contracts (the city's Department of Environmental Protection is responsible for the maintenance and closure plans for its landfills) have ranged up to $50 per cubic yard. That much of an increase in dredging costs (compared with the $4 per cubic yard for historical ocean dumping) represents a big jump in the cost of doing business, and it remains beyond the reach of many smaller facilities.

There are numerous marinas throughout the New York Harbor area, and many of them are in dire straits. A whole group of small marinas in Flushing Bay, on the north side of Queens County with access to the quality sailing and fishing areas of Long Island Sound, have not been able to dredge for years, and many have had to close. Marinas up and down the Hudson River are experiencing similar woes, and the same thing is happening at many other inner harbors, including those in Jamaica Bay in southeast New York Harbor and Mamaroneck in the northeast. These are for the most part private marinas (some are run by local governments) that have provided boating access to the public for many years.

Now that water quality is improving in the harbor area owing to implementation of the Clean Water Act, more people want to go boating again. Many are finding, however, that any local marinas still in business are silted in and severely restricted in capacity. And the high cost of dredging means marinas have to charge higher dockage rates, which are often too high for the average boater to afford. The situation is similar for other small waterfront businesses such as marine maintenance facilities and the smaller goods handlers and shippers.

Under this scenario of high disposal costs, one might ask what the effect will be on the Port's status as a hub port. It can be assumed that necessary funding will continue to be allocated for the federal navigation projects to maintain adequate water depths in shipping channels. However, even the larger corporations that have been able thus far to afford the recent high dredge disposal costs are surely feeling the pinch. As was mentioned at the beginning, the costs of dredging and the reliability of maintaining adequate depth in channels and berths have always been important considerations for the large shipping companies.

The demand for goods in the New York metropolitan area will always be financially attractive, but dredging costs and reliability in maintaining port facilities are at least as important on the other side of the ledger. If the preponderance of dredged material from the Port has to continue to be handled at these greatly increased costs, companies with large dredging needs will certainly take another hard look at other ports. It should also be very clear, regardless of the work of the New York/New Jersey Harbor Estuary Program and other efforts to reduce contaminant inputs, that contaminated sediments in the Port will be with us for a long time. Under the most rosy scenarios it will be at least another twenty years before significant improvement occurs—and these assume that proposed contaminant and sediment reduction efforts will actually take place and be sustained.

It matters little to exporters and importers whether their transportation costs to and from the New York metropolitan area go to the shippers directly plying the Port of New York and New Jersey, or to shippers in other ports and trucking companies that make the first or final leg of the trip. It should matter greatly, though, to New York City and the surrounding area, with its already critical traffic and air pollution problems, which could considerably worsen if much of the cargo coming in and going out of the area is carried by trucks instead of ships. New York City and the two states will do what they can to ensure that the Port survives, both because of the environmental issues and because of the enormous economic issues involved. Whether these and other interests can do enough is still an open question.

It was noted earlier that the Gore agreement recognized the commitments and efforts made by the states of New York and New Jersey in implementing the strategies of the three-party letter for sustaining and maintaining the Port. The PANYNJ was allocated $120 million by Congress, to be split evenly between the two states, to carry out these strategies. New Jersey expended much of that allocation developing land-based disposal actions, largely by underwriting treatment/disposal consortiums to a ceiling of $29 per cubic yard. When the actual costs for some of these projects ended up being much higher, the state had to make up the difference.

The New York allocation was not all expended, and much of what was went to CARP, the Contaminant Assessment and Remediation Program. This program was shared with New Jersey (though New Jersey had a lower cost share) and was supposed to address the causes and sources of contamination in Port sediments. The goal of the program was to collect samples of sediments and various kinds of organism tissues (from benthic organisms to fish to piscivorous birds) from many parts of the harbor estuary and analyze them for contaminant content. From this great store of data the sources of contaminants were supposed to be identified so that they could be dealt with, thereby reducing contaminant input to harbor sediments. The long-term goal was to have cleaner sediments that would not require the more expensive disposal methods that had become necessary. That end would certainly be a desirable one from anyone's standpoint.

Unfortunately, contamination throughout the Port is so widespread and complex that identifying sources can be compared to trying to find the source of the Nile while floating down it in a boat. Sediment contamination is patchy and contamination sources are many, so trying to find a source for any particular contaminant or group of them would require identifying a "trail" from a source to surrounding areas strong enough to overcome the patchiness and the confusing multitude of sources. Estuarine mixing processes and varying river current flows further confound such an effort. Therefore, except in very few cases these trails just do not exist in an estuary like the Port.

In fact, only in a couple of cases were the likely sources for contamination identified, and other than in a very general sense (as in a source being combined sewer outfalls), these were of minor importance. One of these was an agricultural runoff contribution into a tributary far up the Hudson River from New York City, not exactly a major smoking gun. Millions were spent for this effort, and though a lot of scientifically useful information was generated on the extent of contamination in the Port's sediments and organisms, one could question whether this was the best bang for the buck in dealing with the dredging crisis.

So, while New Jersey may have set an optimistically low ceiling in underwriting land disposal projects, thereby expending its $60 million allocation fairly quickly, it nevertheless developed program experience and actually completed several disposal projects. New York put a greater emphasis on the CARP trackdown program as well as on minor efforts in the evaluation of decontamination technologies and in expedited project review procedures. There was little effort put toward identifying disposal alternatives, other than a half-hearted stab at linking the upland remediation program to the dredging review process, in the hope of identifying contaminated upland sites for dredge disposal. Since the effort lacked strong directive from upper levels, there was little follow-through at the upland program division level. One or two minor projects occurred, but overall very little upland disposal occurred in New York until the Jamaica Bay landfills project mentioned earlier, which was not implemented until 2003.

The new best hope for the disposal of Port sediments may now lie in the Keystone State. A pilot project was conducted over the past few years using Port dredged material to remediate a 1.5-mile-long strip-mining area in central Pennsylvania. The project was initiated in 1995 through a public-private partnership between the Pennsylvania Department of Environmental Protection (PA DEP), the New York/New Jersey Clean Ocean and Shore Trust (a bistate legislative group), and Clean Earth Dredging Technologies Inc. The PA DEP has reported that the project is considered to have been very successful (PA DEP website) and is evaluating further authorizations of this kind. Almost 500,000 cubic yards of dredged materials from the Port were mixed with coal and incinerator ash that hardened to form a structural material used to contour and fill the strip mine area. The hardened fill allows rainwater to run off the site instead of mixing with pyritic materials in the raw wound of the mine. (When the mineral pyrite oxidizes, it becomes a key component in the production of acidic mine drainage, one of the biggest sources of pollution in the waterways of the state.) The area was also considered a hazard because of the abandoned mine shafts and steep "high walls" left over from the mining activities, which had occurred before the more modern mining laws came into existence.

The PA DEP estimated that the site was responsible for polluting about 180,000 gallons of water per day as it passed through the area, producing acidic mine drainage. Its report notes that the project brought the area back to grade and thereby removed some of the physical hazards, made surface waters cleaner, and restored natural vegetation and habitat. A key habitat issue has been the restoration of a stream passing through the site, Bark Camp Run, which had once held naturally reproducing trout populations. The PA DEP reports that sections below the site improved sufficiently during the course of the project to enable the return of aquatic insects. Also, despite a statewide advisory for fish consumption of no more than one meal per week, a survey done three years into the project found that overwintered trout in the stream adjacent to the project were suitable for unlimited consumption. Finally, it was reported that five years of monitoring and more than 100,000 analyses found no significant dredge-related contaminants but only those associated with acidic mine drainage that had been present prior to the project's initiation.

Pennsylvania has stated that the reclamation of abandoned mine lands is an environmental priority, with some 5,600 mine sites presenting an array of associated hazards, including underground fires, water-filled surface pits, dangerous high walls, and open mine shafts that tempt the occasional foolhardy would-be explorer. The existing capacity, therefore, for using Port dredged material for these kinds of applications in Pennsylvania, and perhaps eventually other places, is huge.

As usual, there is another side to this project, one that is presented by a Pennsylvania environmental group called Army for a Clean Environment. This group has included an article (Army for a Clean Environment article) on its website written by mining expert Charles Norris, a geologist who works for Geo-Hydro, Inc., a geologic consulting firm. He had completed a preliminary review of the voluminous data that the monitoring effort generated on the materials used and on the environmental field data. Mr. Norris takes a very cautious approach in terms of drawing any conclusions from the data he had reviewed thus far. His main concern is that the mixing and application of various materials and methods throughout the project life span have made it difficult to trace any particular monitoring result to any discrete material mixtures or application methods that were used. For example, he states that there were wide-ranging moisture contents of material shipped from the New Jersey shore-side pretreatment site, where dredged material was initially stabilized with coal fly ash for rail transport to Bark Camp. Also, at different stages of the project different mixtures of coal fly ash and municipal incinerator fly ash were used, with some mixtures apparently not even containing dredged material.

His concern is that this complexity results in an inability to key back any monitoring result to a particular application, thereby providing a procedural alibi for any elevated contaminant findings. Of course, this will be a moot concern if in fact the overall findings indicate, as the PA DEP suggests, not only very little environmental harm but actually significant environmental benefits. Mr. Norris himself notes at the end of the article: "In one sense, the complexity of waste mixing and placement at Bark Camp is very valuable. It clearly demonstrates what can be expected when a permit is issued private entities to manage a site that is authorized for a variety of residual waste streams at various mixes." That can be taken as a criticism of letting this kind of work be done by private entities or, if the PA DEP is right, as a conclusion that even such procedural or operational variabilities can end with acceptable results, which is a good thing.

If the PA DEP determines finally that the processes and materials used for the project are amenable to other sites, this option could be a very important one for disposal of dredged material from the Port. There may be only one remaining factor keeping this option from being a home run, and that is cost. The pilot project required that dredged material be stabilized by the methods discussed earlier, and this will most likely be the case in any similar applications to follow. The long transport distances for this project also increased costs, though some sites closer to the Port may be identified and used in the future.

Stabilization is among the least costly of treatment technologies, and the author has long believed that it is also an environmentally safe method if applied properly. This goes back to graduate work conducted in stabilizing aerospace industry waste with coal-fired power plant fly ash (Lechich and Roethel 1988). The work was an offshoot of CWARP, the Coal Waste Artificial Reef Program, out of the Marine Sciences Research Center (MSRC) at Stony Brook (SUNY). The goal of CWARP was to find mixtures that proved structurally and environmentally suitable for artificial fishing reef blocks placed in the ocean. Dr. Roethel and his graduate students have used mixtures of fly ash with sewage sludge and other materials to conduct long-term studies of the experimental blocks in reefs and other applications. The results of the aerospace waste project and the CWARP projects in general were promising, indicating little potential for adverse environmental effects. Professor Peter Woodhead, a distinguished fisheries biologist at MSRC, was instrumental in conducting or reviewing most of the exposure studies of the blocks in the marine environment.

Economies of scale would probably further reduce the overall costs if a treatment/transport/application scheme becomes established. Still, the option will cost considerably more than historical ocean disposal, and it was the cost of dredging with ocean disposal that was the standard on which many regional marinas and waterfront businesses were established and operated. An order-of-magnitude increase in dredging costs will not fit into the business plans of many of these.

If the Pennsylvania option or a similar scenario is eventually more fully implemented, it will no doubt be a public-private enterprise, with the states of New York and New Jersey intimately involved. The Pennsylvania pilot project incorporated a private entity to receive and initially treat the material in New Jersey. The initially treated material was then transported by (privately owned) rail to the Pennsylvania site. Final treatment and application was done with contractors under the management of the PA DEP. A well-established operation that might result in appreciably lower costs will require close cooperation between New York, New Jersey, Pennsylvania, and a number of private concerns. It will be up to each of the dredging states to determine how to set up eligibility requirements and assess costs for potential dredging projects. There may be a federal role as coordinator/mediator to help pull all these parts together. It will be interesting to see what decisions the states will make regarding subsidizing smaller waterfront businesses, including the marinas.

Treatment technologies that remove contamination from sediments, mostly based on high-temperature incineration or chemical methods, result in materials that can be more easily reused or even sold. However, the costs for all these methods remain prohibitively high for widespread use. But applications for treating hot-spot sediment areas may soon be practical, however, and the manager of Region 2's decontamination program, Eric Stern, remains optimistic. For the vast majority of the Port's contaminated sediments, the stabilization methods with controlled land application may be the predominant alternative. Given the number of brownfields and other contaminated sites around the metropolitan area, and that these procedures have been demonstrated to be environmentally safe if done correctly, stabilization and land application appears to hold much promise. The devil, however, is in the details of how these operations will be coordinated, planned, prioritized, and funded by the states, relevant major cities, PANYNJ, federal agencies, and U.S. Congress.

It's time now to turn our thoughts back to the ocean. A lot of dredged material has been dumped in it and continues to be, not just off New York but around the country and the rest of the world. The revised Green Book procedures have definitely reduced the amount of contamination associated with dredged material that is disposed of in the ocean. The Green Book itself, as well as MPRSA and the Ocean Dumping Regulations, were developed as part of the United States' participation in an international treaty on preventing ocean pollution. The formal name of the treaty is the Convention on the Prevention of Marine Pollution by Dumping of Wastes and Other Matter, but it is more commonly called the London Dumping Convention. At the Mud Dump (now HARS), only Category 1 material can be disposed of. The allowable amount of contamination that is classified under this category may be further reduced in the near future, if not by the time of this publication. The Remediation Material Workgroup (RMW) that EPA has put in place, as was required by the Gore Agreement, has been deliberating on the bioaccumulation thresholds for some time now. These thresholds may be brought down lower, further restricting what is acceptable for ocean disposal.

Under the requirements of the HARS designation rule, material to be disposed of there is supposed to remediate the effects of past disposal. This brings a possibly complicating factor into the tasks of the RMW. The workgroup is trying to arrive at thresholds that will not adversely affect the marine environment, based on relevant risk assessment methods and using toxicological effects data. There is a ceiling for these thresholds, however, that results from the fact that remediation material should be, pretty much by definition, cleaner than the material that it is remediating. Theoretically, a risk-based threshold number can be developed that is considered protective for marine life, yet it may reflect sediment concentrations that are higher than sediments at the site to be remediated. That is, a bioaccumulation tissue threshold value could theoretically pass a sediment with contaminant concentrations as high as or higher than HARS sediments.

The joint Mud Dump closure/HARS designation rule (U.S. EPA 40 CFR Part 228 [FRL-5885-1]) wrote this concern off because HARS sediments and contaminant mixes are so variable they make any site sediment comparisons unworkable. The devil may again be in the details, but there should be an effort to look at all the monitoring data at the HARS (including the early 1990's surveys discussed earlier) and make some judgments about what are the important levels of degradation among the parameters measured (sediment chemistry, toxicity, tissue concentrations). This should then act as a check so that, as the effort grinds toward establishment of updated bioaccumulation tissue thresholds, they reflect contamination levels in sediments that are lower than those determined to be degraded at the ocean site, based on monitoring data.

A SUGGESTION TO ELIMINATE BIOACCUMULATION TESTING

Most likely the risk-based thresholds will be protective, including in comparisons to the disposal site conditions, since the methods are quite conservative.

This leads to a suggestion that should be considered by the agencies and the other stakeholders. EPA and the Corps, as well as the other stakeholders, had not yet reached agreement on updated bioaccumulation tissue thresholds at the time of this writing. A suggestion is that when (and if) they do, the thresholds be applied in the relatively easily described step from benthic tissue to sediment contaminant levels. This would have the effect of eliminating the need for regulatory applicants or federal projects to conduct bioaccumulation tests, the most expensive and time-consuming of the dredged material tests.

This can be done using the Theoretical Bioaccumulation Potential (TBP) formula. There now exists a large body of information developed through the more than ten years of analytical results from the Port and other ports' bioaccumulation testing. With this information and the relationships that can be derived from it, the equation could be applied with a level of confidence that prediction of a benthic tissue concentration from a known sediment concentration could advance from being a theoretical to an actual science. The suggestion hinges on the fact that the relationship works in reverse also. If a benthic tissue concentration (i.e., a risk-based threshold value) is known, a corresponding sediment value can be calculated. This could then be directly applied to sediment chemistry results of a project, obviating the need for bioaccumulation tests.

The benefits of such a testing approach would be considerable, because of the huge cost savings it would mean for ocean disposal permit applicants and federal taxpayers on navigation projects. The author, as well as other experts with whom he has discussed this idea, believe that requiring applicants (and the Corps) to spend hundreds of thousands of dollars for bioaccumulation testing on a project should not continue ad infinitum. If developed appropriately, there should be no concern that this approach would be any less protective of the ocean than if the full suite of bioaccumulation tests were to be carried out. This is because the hardest part of developing thresholds for bioaccumulation tests would already be done, and applied as benthic organism tissue concentrations. If the process of getting to that point is agreed on, it is a very minor step, after completion of the supporting work discussed here, to translate these into sediment concentrations, but that step makes all the difference.

What about the underlying basis for remediating a site in the coastal ocean by capping? The HARS is a 15.7-square-mile area that encompasses the (2.2 square mile) Mud Dump Site and an area around it that has been identified

SCIENCE NOTE: DEVELOPING A NEW TESTING APPROACH

Evaluating all the appropriate data and information to develop a refined TBP approach would require a substantial effort by EPA and Corps scientists. A way this could be done would be to compile all paired sediment chemistry–bioaccumulation tissue data sets and evaluate them for methods consistencies and the presence of associated tissue lipid and sediment TOC (total organic carbon) data. A biostatistician could advise on the appropriate statistical approach(es) toward development of relevant relationships, error ranges, and possibly adjustment factors. It is likely that with the variability in lipid methods used and probably correspondingly in the data, the best that could be achieved may be the development of an acceptably narrow range of lipid values for application. A sensitivity analysis can determine the range of acceptability of the values, possibly linked, as with other variables, to ranges of resulting risk levels. TOC, however, should be a key factor, and an evaluation may well result in bracketing of data around the key values of sediment chemistry, tissue concentrations, and TOC.

Once the relevant data and relationships are established, the approach could be applied nationally, using regional tissue threshold levels. In the Port, the approach could be used to determine Category 1 sediments without the need for bioaccumulation testing, and elsewhere they could be used in whatever comparative decision-making was currently being done with bioaccumulation test results.

to need "remediation" from past disposal activities. It was mentioned earlier that monitoring studies around the MDS had found areas that were contaminated from past dumping. This area was conservatively demarcated to form the boundaries of the HARS and designated as a new dump area but was restricted to only "clean" material that would serve to remediate the prior dumping. For the immediate future, it was intended that what could be considered "clean" would be material determined to be Category 1 by the standards that had been set in place in the Supermemos.

However, as was also mentioned, the Gore Agreement required EPA to conduct a nine-month review of the testing evaluation methods for determining acceptable material for remediation of the HARS. Disposal of accepted dredge projects has continued at the HARS, while the "nine-month" review process has also continued. The environmental groups, notably Clean Ocean Action, have continued to question the present standards, which allow for the dumping of material they do not consider to be "remediation" material.

The science and the monitoring certainly suggest that during some storms some bottom sediments at those depths (about sixty-five feet) will be stirred up and moved around. Whether that mixing will penetrate to below the required one-meter-thick cap of remediation material placed over contaminated areas will have to be answered over the passage of time. There are provisions in the HARS designation that require periodic monitoring and maintenance of the cap. These requirements should not be allowed to slip. The management plan uses a tiered approach, with less costly physical monitoring methods conducted first and more exhaustive methods in upper tiers applied only if triggered by findings in the lower tiers. Bathymetric surveys are tier-one methods that can detect whether there has been significant erosion on mounds at the HARS, and more expensive chemical and biological surveys are conducted only if triggered by the bathymetry results. Given the variables associated with these methods, and as a general check, tier-two chemical surveys should be also be conducted at least every few years, regardless of the tier-one results.

One thing is fairly certain: if the contaminant load put into an ocean area is significantly reduced, there will be a corresponding improvement in the conditions. That was learned from the NOAA 12-mile Sewage Sludge site study. Also, if cleaner material is placed on more contaminated material, it will have beneficial effects regardless of its ability to strictly sequester it by capping. Given a limited mass of contaminated material, such as exists at the HARS, a mass of cleaner material added to it will dilute the environmental risk of the contaminants. Dilution is not a solution for pollution, but this case is the inverse to that venerable tenet. Adding a contaminant stream to a cleaner receiving water (diluting it) will pollute the receiving water, under almost any conceivable conditions. In this case, however, the potential for effects to marine biota from sediment contaminants is a direct function of the contaminant concentrations in the sediments. Adding cleaner sediments to the existing mounds, even if they do only a semi-adequate job of actually containing the more contaminated sediments, will reduce the contaminant concentrations by their addition and mixing.

Reducing those concentrations can substantially reduce the potential for uptake by benthic organisms and on up the food chain. That is because though there are significant variabilities involved, it can be assumed generally that uptake rates are slower and ultimate accumulations of contaminants in organisms lower from sediments that have lesser concentrations of those contaminants. So it can be expected that placing cleaner material at the HARS will have beneficial effects. Of course, the old conundrum of how clean is "clean" is always a factor. From a mass balance standpoint, placing any contamination into the ocean has the potential for buildup and adverse effects.

Since passage of the Clean Water Act in 1972, the waters around the nation have been improving, sometimes to a remarkable degree. Not only are rivers no longer catching on fire; they are improving to the point that large fish runs are coming back to rivers that have not seen them in many decades. This includes the Hudson River, where shad, striped bass, and other fisheries are rejuvenating. Even in the Port of New York, the water is getting so much cleaner that an old problem, from a less contaminated era, has returned. The water is now clean enough for marine wood-boring organisms to have returned to the Port, and their activities are causing extensive deterioration of wood docks and piers. Many such structures are being sheathed with vinyl or concrete to extend their life span, and many new structures are being made of nonwood materials. (No jokes, please, about the good old days of polluted waters.)

The waters are cleaner, but the sediments in many waters are still contaminated from the earlier polluted water conditions, and sediments do not get cleaner as quickly by turning off pollution sources. Many private companies are coming up with innovative ways of treating contaminated sediments, because the incentive is there for developing a truly practical and low-cost method. They know that a huge potential market is there, driven by governments and industry that need to clean up contaminated sediments in many places around the country and the world.

It is clear that our coastal ocean is under threat from pollution, caused mainly by the concentration of human populations along our shorelines but also coming from water and air pollution originating far inland. A majority of the U.S. population lives close to a shoreline, and the trend is continuing as the coastal-area populations grow faster than those in the rest of the country. It is difficult to imagine how the coastal marine environment can improve, considering the mass of humanity right on its borders. Technology can help by decreasing pollution from air sources and providing better treatment methods for sewage and other wastes. But wherever there is a large city, there will be contamination in varying amounts from a number of sources. New York, like many other cities, still has CSOs (combined sewer outfalls) from storm and sewage systems that result in overflow of sewage directly into the Port whenever heavy rains overwhelm the storage/treatment system. Even when operating under more normal conditions, storm sewers bring large amounts of street oils and greases and other contaminants into Port waters. And there is always surreptitious dumping of contaminated waste into sewers or elsewhere wherever there is a concentration of industry. Many of these wastes, in addition to those from permitted industrial and other discharges, end up in Port waters and then go out to the ocean with the tides. This occurs at every industrial and heavily populated port city along the world's coastlines.

Many scientists believe that the dolphin deaths that spurred the marine mammal studies mentioned previously are related to the pervasive contamination of our coastal waters. All kinds of contaminants are detected in the fat and flesh of marine mammals, and it is postulated that these may be having an immunosuppressive effect. The actual diseases or maladies that cause death, or perhaps disorientation in the case of strandings, may be different, but they can all be caused by a general suppression of the animals' immune systems. Causes and effects in these cases are difficult to prove, but if widespread coastal ocean pollution is a source of these problems, what other environmental effects are being produced that we have not observed yet? These distressed species may be the marine version of the canary in the coal mine.

The subject of whale and dolphin beach strandings is a particularly interesting mystery because of the obvious behavioral aspects. There have been all kinds of theories proposed as to why marine mammals beach themselves, usually leading to their death. These theories include periodic anomalies in Earth's magnetic field (which are believed to throw off the animals' internal navigation system), the contamination issue mentioned above, some type of ear parasite (a variation on this is the desire to rub off skin parasites in the shallows), or the suggestion that the gradual depth changes of shallowing water confuses them.

The shallowing-water theory in particular appears to the author to be somewhat contradictory. How can animals with sensory systems so finely tuned as to be able to utilize Earth's weak magnetic field be so confused by gradually shallowing waters? And some species of whales routinely use shallow coves for nursery and rubbing areas with no observed problems. It would most likely have been a considerable evolutionary disadvantage if this behavior even occasionally resulted in stranding and almost certain death, so this theory seems implausible also.

Another, more recently voiced theory is that high-intensity sonar used by the U.S. Navy and other navies is damaging the inner ears of marine mammals and thus their navigation systems. The European Parliament called on member nations in October 2004 to suspend the use of high-intensity sonar during naval maneuvers until research determines whether the loud sounds are leading to the deaths of whales and other sea creatures. In recent years, there have been mass strandings of whales observed following naval maneuvers in the Virgin Islands, the Bahamas, and a few other places around the world. The U.S. Navy has acknowledged that its sonar was responsible for the strandings of seventeen whales in the Bahamas in 2000 but says that it did not play a role in other strandings, and that sonar can be used safely. In July 2004, the International Whaling Commission found that there is compelling evidence that high-intensity sonar has been leading to mass strandings.

AFTERWORD

EPA is nationally still committed to "stakeholder" regulatory negotiations as in the Common Sense Initiative and project XL. Some of these have shown successes, but there has also been criticism that the balance has been tipped more toward regulatory flexibility than toward environmental protection. Command and control environmental regulation is now seen as a dinosaur, not in keeping with the times. But if EPA or other agencies have to negotiate every new controlling regulation before it even gets officially proposed, how is that going to decrease the overall pollutant load already in the oceans, atmosphere, and our food sources? If a new scientific finding is made, for example, that a contaminant at a certain level is harmful to people or the environment, compliance with a new standard set below that level first has to be negotiated with industry. If the new standard is onerous to industry, it will do everything possible to stall, modify, or otherwise discourage the standard being set.

Industry groups have an advance opportunity to stifle a new standard under the "stakeholder" approach, because they can get a crack at it in committee before it ever gets issued for public review. That is a big advantage compared with procedures followed under the traditional regulatory system. If an agency has sufficient confidence in and support for the science behind a new finding, then a standard that reflects that finding should be proposed to all the public as soon as possible. The agency should be able to do this without having to first get approval by a committee that includes special interests with the ability to bottle up change indefinitely.

In late 2004 (as this writing was nearing completion), an air pollution issue regarding mercury became a major concern within the environmental community and among fishermen. The problem of mercury in ocean and fresh water fish has been a concern for a long time, but recent events in the regulation of mercury air pollution have raised new fears. There is a proposal by the EPA under the Bush administration, part of its "Clear Skies" initiative, to revise the 1990 requirements of the Clean Air Act (CAA) for coal-burning power plants. The effects of the revisions would, in the view of some environmental experts, allow for the release of more mercury into the air than would be the case if the CAA original provisions are applied. The CAA required that maximum achievable control technology (MACT) measures be enforced, which meant that allowable releases of mercury had to be within the average measured from the twelve best-performing facilities. This followed an extensive review by EPA at the time that found that of sixty-seven air toxins, mercury was the one of most concern. The MACT approach would have reduced mercury releases by 30 to 35 percent, with some estimates quite higher, by 2007.

According to the National Wildlife Federation (NWF website), the new proposal could allow up to six times those releases under a "cap and trade" program for the energy industry that was developed by "stakeholder" input (sound familiar?). The current EPA administrator apparently feels no concern that under previous EPA administrations a cap and trade approach was considered appropriate only for nontoxic air pollutants such as particulates. New Jersey has already decided that it will stay with the MACT requirements, and other states may follow. An article in the April 2004 issue of *Field & Stream* also discussed this proposal, lamenting that a president who purports to care about sportsmen would promote changes that will likely increase the mercury levels in the fish they catch and bring home to their families. Experts are seeing more people with illnesses that are traced to high mercury levels, and some have stated that whatever specific effects mercury has, and there are many, mercury also generally undermines the overall functioning of the body. Mercury air pollution has direct effects on mercury concentrations in the marine food chain, besides the food chains of inland lakes and streams, through deposition from air to water. Mercury in tuna and other high-trophic-level predators has been a concern for some time, and this concern could well become worse and more widespread to other species.

After the 2004 presidential election, there was a sense of foreboding among the environmental community. There is little doubt that President Bush will push the clean air revisions pertaining to mercury, and given the Republican majority in the House and Senate, he may succeed. In a November 8, 2004, *New York Times* article, Felicity Barringer and Michael Janofsky write that with the elections over, Congress and the Bush administration are moving ahead with ambitious environmental agendas that include revamping signature laws on air pollution and endangered species as well as changes that would open the Arctic National Wildlife Refuge to energy exploration. "The election is a validation of our philosophy and agenda," EPA administrator Michael O. Leavitt said in an interview. "We will make more progress in less time while maintaining economic competitiveness for the country. That is my mission." Phil Clapp, president of the National Environmental Trust, nevertheless warned the White House and congressional leadership against pushing further. He noted that, while nationally the environment was a sleeper issue in the 2004 election, there would be a cost to rolling back environmental laws, since the majority of Americans believe in them. Bush administration officials said after their victory that among the first measures moving toward enactment will be those that govern air pollution levels, specifically, Clear Skies.

Another top priority of powerful congressional Republicans is the thirty-one-year-old Endangered Species Act. Representative Richard W. Pombo of California, chairman of the Committee on Resources, has tried to make it harder for government scientists to make a determination that a species is endangered, and wants to cut back the amount of critical habitat required for such species. The Global Amphibian Assessment study, conducted over three years by more than five hundred scientists (to be published soon in the journal *Science*) concludes that over 1,856 amphibian species are threatened with extinction, which represents 32 percent of all amphibian species. It concludes that the same threat is posed to 23 percent of all mammal species and 12 percent of bird species. The amphibian losses have been widely feared to be the canary in the mine for general environmental degradation around the globe.

Also expected to be pushed is the energy bill that would open two thousand acres of the Arctic National Wildlife Refuge for energy exploration. A third priority, according to Mr. Pombo, is a package of legislation dealing with ocean resources, including issues relating to management of commercial and sport fisheries and the protection of endangered marine mammals.

Given the current situation with the war on terrorism and the concurrent increased ease in citing national security issues, it is unlikely that there will be any finding accepted by the Bush administration that naval sonar has deleterious effects on marine mammals. Expanding the prediction further to marine habitat—related issues, it is unlikely that the past decade's movement toward establishment of marine sanctuaries as a way of rejuvenating fish stocks and protecting unique populations will be much advanced. Marine sanctuaries were first proposed after a devastating oil spill in 1969 blackened the coastline of southern California and killed countless marine creatures, after which the ocean's huge size and currents no longer seemed guarantees to keep it safe from any lasting human damage. Three years later, Congress responded to the oil spill and other accounts of toxic dumping with new environmental laws to better regulate ocean dumping and protect endangered marine animals (MPRSA). Marine areas were identified for their biodiversity, ecological integrity, and cultural legacy, and could now receive protection through the National Marine Sanctuary Program. The program, administered by NOAA, was created in Title III of MPRSA.

These sanctuaries are a type of "marine protected areas" that are created for many purposes worldwide, including the above purposes but also for propagation of fisheries. In new national ocean policy written by the U.S. Commission on Ocean Policy in its Final Report to the President and Congress on September 20, 2004, "An Ocean Blueprint for the 21st Century," marine protected areas were lauded as promising for the protection of marine fisheries and diversity. The study and report were required by the Oceans Act of 2000. Speaking on the overall goals and findings of the study, commission chairman Admiral James D. Watkins, USN (ret.), said, "It's no secret that our nation's oceans and coasts are in serious trouble. It's also clear that as a nation we must rise to the challenge today to reverse the damage to our oceans and change the way we manage them before it is too late. Rising to this challenge will not be easy. It will require strong leadership from the President and Congress, great political will, new fiscal investment, and strong public support. But, in the long run, all Americans will benefit."

Based on the commission's analysis, the total cost to implement the new ocean policy would be $1.5 billion the first year, and the annual cost rises to $3.9 billion in the out years. It was stressed that the proposed funding levels would be a modest investment to protect one of the nation's greatest natural and economic resources. Of the nation's 259 major fish stocks, 20 percent are either overfished or experiencing overfishing.

The commission's final report (U.S. Commission on Ocean Policy 2004) contains 212 recommendations, targeted to Congress, the executive branch, federal agencies, regional bodies, and international affairs. Most of the recommendations are geared to encourage better protection of marine resources in a myriad of areas, and many suggest the creation of special boards or committees to fine-tune and oversee the recommendations. Though there are a number of recommendations regarding increased research into marine environmental issues in general, there is no specific one regarding the use of naval sonar and its effects on marine mammals. A main recommendation to the Corps of Engineers is that its "selection of the least-cost disposal option for dredging projects reflect a more accurate accounting of the full range of economic and environmental costs and benefits for options that reuse dredged materials, as well as for other disposal methods."

President Bush formally responded to the report on December 17, 2004, as was required by the Oceans Act (White House 2004). The White House response, a forty-page "U.S. Ocean Action Plan," was written by the Interagency Ocean Policy Group (IOPG), a group described as consisting of senior federal officials drawn from eight cabinet and independent agencies with relevant policy experience in ocean and coastal resources. The plan lays out nine action items that frame the administration's response. The first item on the agenda has already been completed: President Bush issued an executive order to create a cabinet-level Committee on Ocean Policy. The committee will coordinate the activities of executive branch departments and agencies regarding ocean-related matters and will recommend and implement more detailed responses to the sweeping "Ocean Blueprint" report.

The executive plan recognizes that integration of federal effort is a critical need for ocean policy, since more than twenty federal agencies administer over 140 federal laws related to the oceans, coasts, and Great Lakes. The plan's steps toward this integration include convening the policy group's first meeting early in 2005 to develop an eighteen-month work plan to address the commission's recommendations, including actions on governance principles, filling gaps in legislative authority, and streamlining unnecessary overlapping authorities. One of these actions will be for the Joint Subcommittee on Ocean Science and Technology, part of the National Science and Technology Council, to develop a "Framework for an Ocean Research Priorities Plan and Implementation Strategy" by the end of 2006.

Other proposed actions include creating a global earth observation network (which would include integrated ocean observations); building and deploying new state-of-the-art research and survey vessels; creating a national water quality monitoring network; providing for more efficient ocean and coastal mapping activities; implementing new legislation relating ocean policy to human health, harmful algal blooms, and hypoxia; and improving public education on oceans.

Reaction to the administration's plan has been generally positive, with a wait-and-see attitude regarding its implementation. EDF sees the plan as a positive step and has urged President Bush to follow through and possibly go further, noting he has the "opportunity to do for oceans what Teddy Roosevelt did for land." Representative Sam Farr (D-Calif.), cochair of the House Oceans Caucus, has said he hopes this will not be the only decisive action the president takes on oceans: "The real question today is, how much muscle is the White House willing to put behind their Ocean Action Plan? Are they willing to tackle the more difficult but critical governance issues?" It will have to be seen whether President Bush takes these environmental issues regarding oceans seriously.

It is apparent that many people are concerned nowadays about the environment and want to do the right things, and this has resulted in progress. It is also clear, though, people are not very willing to do the right thing if it is not convenient or if it has a cost. When recycling methods were made more convenient (i.e., when combining different kinds of recyclables was allowed), more people recycled. The ingenuity of talented people in the many fields that pertain to reducing water and air pollution, hazardous wastes, and the huge quantities of solid waste in our society need be brought to bear to help solve these problems in ways that the American public will embrace.

Technology alone cannot solve all these problems, though—some level of personal sacrifice or willingness to change one's lifestyle will be necessary. The widespread use of plastic shopping bags is an example of a harmful practice that many European countries have addressed by shoppers bringing their own bags to stores. Unfortunately, our wholesale use of heavy duty plastic bags for garbage disposal will ensure that our landfilled waste will stay fresh for future generations to enjoy for many years to come. When remnants and particles of plastics do get into the environment, they usually have bad effects, such as when marine life mistake them for food and clog their digestive systems—often leading to death.

There must be a better job done not only of environmentally educating the public but also of promoting pride in and concern for the environment in people's everyday lives. Individual actions can have more widespread effects than just the direct result of such actions, when other people are made more aware by them. This holds true in the positive sense as well as the negative. There are many outreach programs targeted to schoolchildren that promote a concern for the environment, and they should be expanded and supported.

On the other hand, individuals or groups that despoil the environment are responsible not only for whatever direct damage they cause; they should also be censured for the bad example they set. In everyday life, we observe people who deliberately throw garbage onto streets or into waters, in plain view of children who are watching.

We are also subjected to the insult of far right-wing commentators who try to portray a concern for the environment as left-wing adventurism that will destroy the economy. As they rail against the environmental laws that have cleaned up our waters and air and conserved wild lands, one wonders what cognitive defect could be responsible for their inability to see that their own children, much less themselves, are better off because of them. Or perhaps they believe they can insulate themselves and their families in air-filtered homes and survive on bottled water and purified food, and venture outside under the protection of ultraviolet-blocking devices. Judging by the otherwise normal life skills portrayed by people of this mind-set, it is doubtful they actually believe half the nonsense they spew. This would, however, relegate them to the even lower status of myopic mercenaries for the powerful, with adopted agendas contrary not only to the good of society as a whole but also to that of these confused souls themselves.

It will take practical-minded people to scorn this kind of menial thinking, because it is not just another legitimate viewpoint in a highly divergent society, as some would say. It is an anachronism that we can simply no longer afford.

Whether such a practical-minded, environmentally sound sea change in thinking and attitude takes a strong enough hold over a large enough portion of the population to effect real change remains to be seen. It will probably be a necessity, however, if we are going to be able to enact and implement better controls over the remaining major sources of pollution, including those that degrade our oceans and coastal waters. It may be the only way that we will be able to overcome insistent cries from industry about profit-killing costs of pollution controls, and politicians' fears of legislating policy that would enable more environmentally friendly ways of going about our business.

REFERENCES

Army for a Clean Environment article website. armyforacleanenvironment.org/bark .html.

Kociba, R. J., and B. A. Schwetz. 1982. Toxicity of 2,3,7,8-tetrachlorodibenzo-*p*-dioxin (TCDD). *Drug Metab. Rev.* 13:387–406.

Lechich, A. F. 1998. Development of Bioaccumulation Guidance for Dredged Material Evaluations in EPA Region 2. National Sediment Bioaccumulation Conference, Proceedings. US EPA, Office of Water. EPA 823-R-98–002. February.

Lechich, A. F., and F. J. Roethel. 1988. Marine disposal of stabilized metal processing waste. *Journal of Water Pollution Control Federation,* Washington, D.C. January. Graduate research conducted for completion of M.S. in Marine Environmental Science, Marine Sciences Research Center, SUNY, Stony Brook.

L. Don Leet and Sheldon Judson. 1965. *Physical Geology.* Third Edition. Prentice Hall: Englewood Cliffs, N.J.

Long, E. R., and L. G. Morgan. 1991. The Potential for Biological Effects of Sediment-Sorbed Contaminants Tested in the National Status and Trends Program. NOAA Technical Memorandum, National Oceanic and Atmospheric Administration, NOS OMA 52. Seattle: NOAA.

Newell, A. J., D. W. Johnson, and L. K. Allen. 1987. Niagara River Biota Contamination Project: Fish Flesh Criteria for Piscivorous Wildlife. New York State Department of Environmental Conservation, Division of Fish & Wildlife. Technical Report 87–3.

NWF (National Wildlife Federation) website. www.nwf.org/mercury. Go to Washington, D.C., Office button for the Clean the Rain program and mercury in power plants proposal.

PA DEP (Pennsylvania Department of Environmental Protection) website. www.dep .state.pa.us.

Pruell, R. J., N. I. Rubinstein, B. K. Taplin, J. A. LiVolsi, and R. D. Bowen. 1993. Accumulation of polychlorinated organic contaminants from sediment by three marine species. *Arch. Environ. Contam. Toxicol.* 24:290–298.

R. J. Pruell, B. K. Taplin, D. G. McGovern, R. McKinsley, and S. B. Norton. 2000. Organic contaminant distributions in sediments, polychaetes (*Nereis virens*) and American lobster (*Homarus americanus*) from a laboratory food chain experiment. *Mar. Envir. Res.* 49:19–36.

U.S. Army Corps of Engineers, New York District, and U.S. EPA, Region 2. 1992. Guidance for Performing Tests on Dredged Material Proposed for Ocean Disposal. December.

U.S. Commission on Ocean Policy. 2004. An Ocean Blueprint for the 21st Century," Final Report. September 20. www.oceancommission.gov/documents/welcome.html.

U.S. EPA. 1997. Ocean Dumping Regulations. Title 40 CFR Parts 220–230.

U.S. EPA. 1993. Interim Report on the Assessment of 2,3,7,8-Tetrachlorodibenzo-*p*-dioxin Risk to Aquatic Life and Associated Wildlife. Revised January. EPA/600/XXXX. Washington, D.C.: Office of Research and Development.

U.S. EPA 40 CFR Part 228 [FRL-5885-1]. Simultaneous De-designation and Termination of the Mud Dump Site and Designation of the Historic Area Remediation Site AGENCY: Environmental Protection Agency. www.epa.gov/Region2/water/dredge/harsrule.pdf

U.S. EPA website. http://www.epa.gov/ORD/NRMRL/std/mtb/csi.htm. EPA's website has changed over the past few years, but this site, as of April 2004, contains several-years-old material on the Common Sense Initiative and reinventing regulation.

U.S. EPA and U.S. Army Corps of Engineers. 1991. Evaluation of Dredged Material Proposed for Ocean Disposal, Testing Manual. EPA-503/8–91/001. February.

White House. 2004. U.S. Ocean Action Plan. Response to U.S. Commission on Ocean Policy Final Report, An Ocean Blueprint for the 21st Century. www.ocean.ceq.gov/.

food chain, 25, 63, 118; Framework for
an Ocean Research Priorities Plan and
Implementation Strategy, 167; House
Oceans Caucus, 168; pollution, 56, 156,
161, 164; pits, 54; White House Ocean
(U.S.) Action Plan, 167
oil spill, 166
orbital velocity, 32
organic material, 112
oxygenated (water), 41, 42, 122

PAH (polynuclear aromatic hydrocarbon),
132, 144
PADEP (Pennsylvania Department of
Environmental Protection) PA mines
project, 152, 153, 154, 155
Pallone, Congressman Frank R. 84, 125
pH, 8
phytoplankton, 91, 94
pits. *See* borrow pits
Port Newark/Elizabeth, 16, 59, 60, 65, 66,
71, 75, 80, 81, 82, 83, 87, 88, 89, 99, 110,
113, 119, 137, 148
prey, 67, 68
Pruell, Richard, 23, 24, 26, 83

quadropod, 33

Reg-neg, 75, 138, 163
remediation, 49, 82, 109, 147, 152, 156,
158; Remediation Material Workgroup,
156; sites, 108, 149

salinity, 62
sea bass, 118
seawater, 33, 40, 41, 42, 62
sediment, coastal, 48, 63, 109, 114;
control, 86; organic (or silty), 43, 48,
112, 114, 127; reference; 40, 51, 111, 112,
116; sandy, 63, 124; surficial, 62, 82, 122;
suspended, 43, 57, 90, 91, 114; test (or
project), 52, 61, 62, 91, 111, 140; /water
interface, 41, 42

sewage sludge, 14, 28, 29, 47, 48, 50, 137,
153
sonar high-energy, 161, 165, 167; side
scan, 32
shad, 160
shelf (Continental) break, 48
SMIRF (Sediment Management Issue
Resolution Forum), 79
static (test), 40, 41, 102
statistical (methods), 86, 158
stakeholder(s), 16, 76, 79, 138, 157, 163,
164
stranding(s), 161, 162
Striped bass, 160

TEQ (Toxic Equivalents, for dioxin), 92,
93, 94
Theoretical Bioaccumulation Potential
(TBP), 127, 157, 158
tidal, 3, 15, 62
tide(s), 127, 161
tomalley, 83
total organic carbon (TOC), 127, 158
toxicity, 65, 66, 138
toxicity test, 21, 40, 43, 58, 59, 62, 63, 64,
91, 100, 105, 109, 11, 112, 130, 156
Toxicity Characteristic Leaching Procedure (TCLP test), 129
trophic, 106, 108, 138, 164

Wakeman, Tom, 140
Waterways Experiment Station (WES,
U.S. Army Corps of Engineers), 10, 39
whale(s), 70, 161, 162
woodburning (at sea), 13, 14
Woodhead, Peter, 154

XL (Project), 138, 163

Zambrano, John, 108
Zipf, Cindy, 14
zooplankton, 91, 94
ZSF (Zone of Siting Feasibility), 30, 32